T0141048

INTRODUCTION TO GAME THEORY

INTRODUCTION TO GAME THEORY

A BEHAVIORAL APPROACH

Kenneth C. Williams
MICHIGAN STATE UNIVERSITY

NEW YORK   OXFORD
OXFORD UNIVERSITY PRESS

Oxford University Press, Inc., publishes works that further Oxford University's
objective of excellence in research, scholarship, and education.

Oxford New York
Auckland Cape Town Dar es Salaam Hong Kong Karachi
Kuala Lumpur Madrid Melbourne Mexico City Nairobi
New Delhi Shanghai Taipei Toronto

With offices in
Argentina Austria Brazil Chile Czech Republic France Greece
Guatemala Hungary Italy Japan Poland Portugal Singapore
South Korea Switzerland Thailand Turkey Ukraine Vietnam

Copyright © 2013 by Oxford University Press, Inc.

For titles covered by Section 112 of the US Higher Education
Opportunity Act, please visit www.oup.com/us/he for the
latest information about pricing and alternate formats.

Published by Oxford University Press, Inc.
198 Madison Avenue, New York, New York 10016
http://www.oup.com

Oxford is a registered trademark of Oxford University Press

All rights reserved. No part of this publication may be reproduced,
stored in a retrieval system, or transmitted, in any form or by any means,
electronic, mechanical, photocopying, recording, or otherwise,
without the prior permission of Oxford University Press.

Library of Congress Cataloging-in-Publication Data

Williams, Kenneth C.
Introduction to game theory: a behavioral approach / Kenneth C. Williams.
   p. cm.
Includes bibliographical references and index.
ISBN 978-0-19-983739-7
1. Game theory. I. Title.
HB144.W55 2013
519.3—dc23   2012005801

Printing number: 9 8 7 6 5 4 3 2 1

Printed in the United States of America
on acid-free paper

BRIEF CONTENTS

CONTENTS

PREFACE

Behavioral game theory uses psychological reasoning to explain behavior within a game theory model and tests this behavior using laboratory experiments. As will become evident from this book, I truly believe that many (although not all) social science research problems lend themselves to an axiomatic treatment and empirical testing. I understand that research methods are an acquired taste and many people do not share my view, so this book is an attempt to expose behavioral game theory to a wider audience.

I believe (as do others) that the reason that game theory and laboratory experiments are not commonplace in the social sciences is because they have a reputation of being difficult because of the math they involve. The point of this book is to show that math should not be an obstacle, because the basic game theory material is fairly easy to learn and the math is often not the key component of the concepts. Rather, game theory is a way of thinking logically about a problem and formulating research questions based on theory. This method forces a researcher to strip down a problem to its core elements to expose its internal machinery. This allows us to carefully inspect how the parts of the problem fit and operate together to produce some outcome. Experiments can be used to observe whether the parts work in the way that they were predicted to work. In my opinion, this approach is important, especially for young scholars, because it forces them to dissect and systematically study problems.

Behavioral game theory requires a knowledge of basic game theory and basic experimental methods. Outside of psychology, few students in the social sciences learn about experimental methods, and outside of economics, few students learn about game theory. The blending of the two areas provides a unique way to study human decision-making. This book attempts to teach the basic material in a nontechnical way that is accessible to a lay audience. I find that repetitive learning is useful with this material, since people who are being exposed to its ideas for the first time can easily become confused. Consequently, there are some concepts that I attempt to "hammer home" by repeating them several times in slightly different ways in the hope that the various explanations might make the concept make more sense.

I consider the problem sets at the end of each chapter to be an integral part of the overall book. The problem sets are designed so that the reader can get a hands-on feel of how game theory models and experiments actually work, since they require the reader to construct models and think about designs to test them. While some of the problems are more challenging than others, I hope that if a solution is not apparent, readers will take this approach: think about how the problem should have been asked, and then solve that problem. This book is a result of courses that I have taught on research methods, experimental methods, game theory, and behavioral game theory over the years. As all instructors know, some concepts work perfectly in the classroom and others are a disaster. There is a learning curve in developing a course, in which the more the course is taught, the more disastrous material can be removed and replaced with more appropriate material. I have taken courses from Peter Diamond and Drew Fudenberg, so I have seen minor flubs in the classroom from the best. My course, as well as this book, has undergone this transformation, and I must apologize to students who took my early courses on this topic and those students and reviewers who read an early draft of this manuscript.

Since this book is mostly concerned with displaying the merits of laboratory experiments in the study of human behavior, one of the biggest problems I encountered was selecting experiments to be included in the various sections. My goal was to select a diverse set of experiments so that the reader could be exposed to a wide range of different experimental designs. Consequently, some of the experiments I have included may seem odd or peripheral to a particular area. My criteria for selecting the experiments were that the experiment had to be easy to understand, and that the experiments should be different from each other in terms of design as much as possible. Although I know that I have excluded many experiments that should have been included, I hope that my selection provides the readers with exposure to a diverse range of experiments.

Some people criticize game theory for its simplicity, especially regarding the rationality assumption. My view of rationality—and a view that I have emphasized in this book—is that the assumptions of rationality are a modeling technique that allows us to model fundamental human behavior using math functions. Game theory models are concerned with institutions, so we have to have a way to model decision-making behavior in order to study the impact of institutions on behavior. Rationality is not a theory about how real people behave when they make decisions in a consistent manner, but a modeling technique that allows us to mimic a fundamental aspect of human decision choice using a mathematical function.

I like to think of this book as a collaborative effort with my students and the many anonymous reviewers who took the time to write very detailed comments. I am very thankful to the publisher for selecting a very diverse set of reviewers so that I received a diverse range of comments. I attempted to be meticulous about responding to every comment, since I knew that if someone felt compelled to write a comment on an issue, then I needed to respond to that issue. Even those comments that I initially felt were misguided ultimately became very useful in helping me clarify my discussion. There were two reviewers in particular who spent a considerable amount of time on early drafts of this manuscript, giving me very detailed constructive and critical comments and pointing out the errors (of which there were many) that I had made. You know who you are, and I thank you.

I would also like to take the chance here to say a few goodbyes, the first being to one of my advisors, Mel Hinich. I have met a few true geniuses in my lifetime, and I would have to put Mel near the top of that list. The reason that he surpassed others was his command of knowledge in an amazingly large number of diverse fields. Looking at the breadth of his research, it is almost impossible to classify him. He understood that true knowledge came from the integration of different fields of study. He once told me that almost all researchers hide in their own research holes and fail to look up and see what other researchers are doing in their holes. He felt that some of these holes could be connected, and that this connection was the only way that we could discover truth. I can take solace in the fact that his influence on me was common knowledge between us. In a similar vein, I would like to pay homage to Richard McKelvey, with whom I had the honor of working on an experimental project as a graduate student. I recall that I was supposed to conduct the experiment, and he was rather perturbed and felt that I should not be doing it, since I knew the theory behind the experiment. This was a valuable lesson about the seriousness of experimental controls. Lastly, I would like to pay my respects to Haywood Alker, whose narratives still mystify me. I would also be remiss if I did not thank my other advisors, especially my head advisor Peter Ordeshook, and the also important Mat McCubbins (my Austin home), Benjamin Page, and Brian Roberts. Also around helping out during my grad school days were Gary Cox (my other Austin home), Mike Munger, Terry Sullivan, Tom Schwartz, Jim Sidanius, and Gavin Duffy (who introduced me to artificial intelligence).

I also want to thank my editor Jennifer Carpenter and her assistant Maegan Sherlock for helping to develop this project, and Keith Faivre for editing assistance. It's a luxury to write without having to worry about page limits! Of course, I need to thank Becky Morton, who over the years has forced me to think "a lot" about "the proper way" to conduct experiments. Thanks also to Sugato Dasgupta for fixing some mistakes for me, and to Rick Wilson for advice about in-class experiments from which I "borrowed" some designs. I would also like to thank the following reviewers for their feedback: Philip Arena, University at Buffalo; Anna Bassi, University of North Carolina at Chapel Hill; John Morgan, University of California, Berkeley; and Alex Weisiger, University of Pennsylvania. Finally, I would like to thank my family, especially my parents Elijah and Julia and brothers Reggie and Greg, for all of the support they have given me over the years. I would like to extend a special thanks to my wife Marcie, the Soviet historian and lawyer who really doesn't "get" the social science thing but always provides useful comments and even suggested the most common names she knew, Olga and Igor, for the featured players in this book. Finally, I wanted to thank my daughter Katie for her patience in understanding that Dad's games are not really fun games to play.

CHAPTER 1

WHAT IS GAME THEORY?

We can't solve problems by using the same kind of thinking we used when we created them.

Albert Einstein

A. THE GOAL OF THIS BOOK

1. BASEBALL STADIUM MODEL EXAMPLE

This book is about building and testing models. To illustrate what I mean by this statement, consider the following example. After the 1988 baseball season, the roof at Fenway Park Stadium (the baseball stadium for the Boston Red Sox) was torn down behind home plate and replaced with a higher roof to accommodate the creation of luxury corporate box seats.[1] Paul A. Lagace, a professor of aeronautics and astronautics in the Engineering Department at the Massachusetts Institute of Technology, was an avid Red Sox fan and a season ticket holder with a regular seat in center field. During the 1989 season he had noticed that very few home runs were hit to center field. After some consideration, he hypothesized that the new roof section might be the culprit. He was curious about why the roof would affect home run hits to center field, so he had his undergraduate students build a miniature wooden model of Fenway Park Stadium with interchangeable short and tall roof sections behind home plate. The research team put the model stadium in a wind tunnel and varied the roof size to see the impact this would have on the distance of the flight of the ball. They found that the taller roof created a vortex that caused baseballs hit to center field to travel 10 feet less in distance, which accounted for the lack of home run hits to center field. The Red Sox management subsequently redesigned that section of the roof.

This example illustrates what I mean by building and testing models. A model is an abstract representation—not a precise miniature replica—of some real-world situation or event. Events in the real world are often too complex to study because of the interconnected phenomena that produce them. Models allow us both to simplify the real world so that different phenomena can be untangled and studied and to make predictions about

1

phenomena. A test of the model can determine how well the model predicts and how to better understand some phenomena.

2. APPLIED MODELS VS. PURE THEORY

In the previous example, the baseball stadium model did not replicate all aspects of the real baseball stadium (such as the restrooms in the lobby of the stadium), but only the aspects that were needed to test the research question (removable box seats and roof). This type of model is referred to as an applied model, because it is primarily built to be tested. Applied models are constructed to allow researchers to vary the parameters of the model and apply empirical tests to observe other variations that occur.[2] In the baseball example, the miniature baseball stadium was the applied model, the roof section was the parameter that was varied, and the wind tunnel was the empirical test. As opposed to applied models, pure theory models are concerned with more stylized aspects of the real world (although pure theory often becomes applied theory). For the most part, these models are valued for the results that they predict and are not intended to be tested in the sense that applied models are. Although pure theory models provide important insights, this book is concerned only with applied models, or models that are constructed to be tested.

3. APPLIED MODELS AND EMPIRICAL TESTING USING EXPERIMENTS

Unlike the baseball stadium model in which the concern was the interaction of wind currents with the flight of a baseball, in this book we will be concerned with models of social interaction—that is, models that examine the interactions people have with others in a particular institutional setting. Like the wood that was used to replicate a model of the Fenway Park stadium, the building material for the models we will examine in this book are the assumptions used to define game theory. An assumption is simply a statement or proposition that is assumed to be true when conclusions can be logically derived from it. A set of assumptions is used to specify a game theory model and a solution to the model is derived by logical deduction based on the specified assumptions.

An applied model requires some sort of empirical data to test the results it predicts. There are generally two sources of data: field or observational data and experimental data. I will discuss this distinction in more detail later, but this book is mostly concerned with experimental data. Experiments allow for the test of a model in the absence of field data or data that naturally occur in the real world. In an experiment, a researcher is able to set up a controlled environment in which all aspects of the environment can be manipulated and studied. The baseball stadium created a controlled environment for the MIT researchers that allowed them to isolate those factors that impacted the flight of the ball. The experiments that we will study here are like the baseball stadium test in that we will study human behavior within some controlled structure to observe whether behavior conforms to the model's predictions.

Experiments allow for empirical tests of a model by isolating the concepts or variables of interest so that they can be measured and observed. A variable is simply a concept that we can measure, such as the velocity of the wind, the volume of the stadium, the height of the roof, or the flight of a baseball. In our example, the main concern was the distance of the baseball's flight, so parameters in the model focused on isolating this variable and then interacting it with other variables (height of the roof) to determine its impact. In applied models, some variables are fixed, such as the volume of the stadium, while other variables are allowed to vary, such as the height of the roof. An empirical test allows a researcher to specify and observe the interaction of various variables to determine how this interaction impacts outcomes (e.g., Was the flight of the baseball hindered by certain variables?).

In the previous example, the researchers used a wind tunnel and miniature baseballs to test the prediction that the height of the roof impacted the flight of a baseball, and as a result, we learned that taller roofs do, in fact, affect the distance a baseball travels. In this book, we will also be concerned with testing our models, but instead of wind tunnels and baseballs, we will be studying human subjects in experiments that test the assumptions and solutions of game theory models in a controlled environment. As we will see later, controls over the experimental environment are important, because they allow a researcher to recreate a model that is written on paper in the experimental laboratory. That is, we can recreate a mathematical model in a laboratory experiment so that the model can be tested.

Experiments are conducted primarily to establish causality or determine how variables of a model are related by holding a particular variable of a model constant (e.g., wind in the stadium) while varying other variables (e.g., the height of the roof or the removal of box seats). But instead of using a variable like wind, we will examine our models using real people. Using human subjects (people who participate in an experiment), a researcher can populate a model and observe whether real people make decisions that are in accordance with the decisions that the players in the model are predicted to make.

4. A SIMPLE AND NOT VERY GOOD EXPERIMENT

I have thus far used the word "experiment" a lot, so to illustrate what I mean by this term, I will present a very simple example that anyone can conduct. Let us say that you notice that when you shake hands with someone and squeeze really hard, the person with whom you are shaking hands seems to frown, and when you squeeze gently, the person seems to smile. You are puzzled by this behavior and want to determine if there is a relationship between the type of handgrip used when shaking hands and the type of emotion that is evoked. To examine this relationship, you conduct an experiment with two treatment variables (i.e., the two variables you want to compare or vary). In our experiment, the parameter that varies is the type of handgrip. Hence, to conduct the experiment, you must set up two treatments: one in which subjects are only exposed to a hard grip and one in which subjects are only exposed to a gentle grip. Then you can observe the occurrence of smiles and frowns in each treatment and compare the results to examine their relationship.

To conduct the experiment, you determine that you will shake 20 random strangers' hands; in Treatment 1, you will squeeze the first 10 strangers' hands really hard, and in Treatment 2, you will squeeze the next 10 strangers' hands gently.[3] You go out and conduct your treatments on 20 different people and record whether they smile or frown depending on the type of grip used. You also record any other behavior besides a frown or smile (e.g., a neutral facial expression) that you observe. This other type of behavior constitutes a deviation from your prediction of either a smile or a frown. While this is not a very good scientific experiment (since there are no definitions of what constitutes a hard or soft grip or a frown or smile), it is an experiment nevertheless, and it does illustrate the thought process involved with experimental research—that is, the isolation and comparison of variables. Although there are flaws in the test's design, the experiment would provide some (weak) evidence concerning the causal relationship between the type of handgrip and the emotions this gesture evokes.[4]

5. BEHAVIORAL GAME THEORY AND ULTIMATUM BARGAINING

This book takes a behavioral game theory approach that studies game theoretic concepts through experiments and examines how real people make decisions when they populate a particular model. The problem behavioral game theory addresses is that when real people participate in an experiment to test some game theory model, their behavior sometimes deviates from the predicted behavior. Behavioral game theory asserts that some of this deviation can be explained by human emotions that are not assumptions of the model being tested. To illustrate what I mean, consider the ultimatum bargaining game, which is one of the simplest bargaining models in game theory and one that provides a clear-cut subgame perfect equilibrium prediction. I will provide a precise definition of subgame perfect equilibrium prediction later in this book, but for now, just think of it as a solution to a game or what we would expect to happen given the assumptions of a model. In this bargaining game, two players bargain over some fixed amount of money, say 10 dollars. Player 1 can make a proposal for how to divide the 10 dollars and Player 2 can either accept or reject that proposal. If Player 2 accepts the proposed allocation, then both players receive the specified amounts, but if Player 2 rejects the proposal, then both players receive nothing. The equilibrium prediction is for Player 2 to accept any positive amount, so, for example, if Player 1 offers Player 2 one cent and proposes to keep $9.99, the equilibrium prediction is that Player 2 should accept this proposal. But what happens when real people play this game in an experiment? Will Player 2 accept any positive amount that Player 1 offers, such as one cent? The answer is that sometimes Player 2 will and sometimes he or she will not. The model predicts that if people care for money and more money is always preferred to less money, then any positive amount of money is better than none. Hence, the solution indicates that players will be happy with a penny. However, some people may object to getting only a penny as an issue of fairness. A person might ask him or herself, "Why should I take a penny while this other person gets almost 10 dollars?" In this case, a person may feel that nothing is better than a penny—a behavior that does not conform to the prediction of the model.

Zamir (2000, 5), who is one of many scholars who have conducted experiments on this type of bargaining game, reports the insightful comment of one subject after completion of the experiment: "I did not earn any money because all the other players are *stupid!* How can you reject a positive amount of money and prefer to get zero? They just did not understand the game! You should have stopped the experiment and explained it to them."

One of the main ideas that I explore in this book is the following question: Were the other players stupid and unable to figure out the solution to the game, or were these stupid players actually playing and understanding a different game, one not modeled by the researcher and to whose rules their behavior conformed? This different game could be one that involves emotions such as fairness, equality, anger, or revenge. For instance, Player 2 might feel that a 50/50 split would be a fair offer and might reject a lower amount out of concern for issues relating to fairness, anger, and so on. Hence, when real people play these simple games, they often play the game differently than predicted. This unpredicted behavior creates the interesting research question of *why?* Behavioral game theory is concerned with psychological variables that might explain this unpredicted behavior.

6. NEW TECHNOLOGY USED TO DISPROVE AND IMPROVE OLD THEORIES

Experimental tests of game theory models are a relatively new technology, since a systematic experimental approach did not appear until the 1980s (see Appendix 2 for a brief history of game theory experiments). Experiments have redefined the importance of game theory models and allowed for the revision of old theories into new and more interesting theories. Much of this revival has to do with new technology such as computers, which allow more sophisticated experiments to be designed. In turn, the increased level of sophistication allows more precise testing of game theory models, which has resulted in the current emphasis on behavioral game theory.

All sciences follow a natural progression of testing existing theories in which predictions need to be revised as a result of new technology. For instance, it was the prevailing theory in astronomy that planets rotate in the same direction that their host star rotates. However, theories in astronomy have been altered as a result of new technology like the Hubble telescope, which can make observations from a distance of over 12 billion light years. Recently, the United Kingdom's Wide Area Search for Planets (WASP) cast doubt on this theory by discovering a planet one thousand light years away that rotates backward around its host star rather than in the same direction. Astronomers have speculated that a prior collision with another planet caused this odd rotation (Harmon 2009). In fact, scientists recently found 10 Jupiter-sized planets that did not revolve around any host star at all, challenging the standard definition of a planet and possibly causing new definitions to be established. Are massive "free roaming" rocks planets? Scientists have hypothesized that these "rogue" planets were ejected from their orbit by the gravitational pull of larger planets (Bennett 2011). These observations do not prove that the entire theory of orbital motion is invalid; rather, they show that an assumption of the theory is false and demand that the theory be revised accordingly.

Consider another example from oceanography, in which a prediction of a long-existing model was recently disproved by new technology (Bower et al. 2009). For the last 50 years, the prevailing model concerning ocean currents has been a "conveyor belt" model, in which deep cold water from the Labrador Sea flows southward and westward to form a continuous loop with the northwest-bound currents called the Gulf Stream. This continuous loop is supposed to divert cold water from the Labrador Sea to the tropics as it travels south, making that region a little colder and Europe a littler hotter. Figure 1.1 shows the pattern that is predicted by the model.

This model hinged on the assumption that the water from the Labrador Sea was diverted to the south and the west in the direction of the Gulf Stream. However, researchers discovered that the currents of the Labrador Sea actually traveled south and then east, proving that the old model's prediction was wrong. The prevailing model was supported by empirical evidence that was produced by biased measurements and outdated technology. Previous empirical tests of this model used floats submerged in the deep current (located at a depth between 700 and 1500 meters) that researchers would follow to establish the path of the current. After finding the floats, researchers found that the floats did follow the pattern predicted by the model. The problem with this empirical test was that the floats had to return to the surface for researchers to determine their positions, but by the time the floats could be located, their positions had been compromised by the upper currents. New technology allowed the floats to be fitted with a seismic signal that permitted researchers to track their location while still submerged. The researchers placed 76 "new" floats in the Labrador Sea and traced their trajectories. Consequently, they found that only 8 percent of the floats followed the conveyor belt model, while many more followed an eastward path. Hence, this discovery invalidated a prediction that had been believed to be true for 50 years.

This example shows that models are powerful tools for examining naturally occurring situations, since they allow us to simplify nature in order to systematically determine how one variable impacts another variable. Empirical observations then allow us to determine if our variables behave in the manner that was predicted by the model. When new ways to

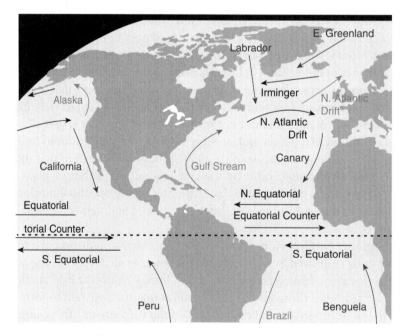

FIGURE 1.1   Gulf Stream currents model

make measurements for empirical observations arise, then standard models may need to be revised. Behavioral game theory is premised on the notions that an emotional type of behavior can be induced and measured within the framework of a game theory model, and that these same emotions can also be tested empirically. Game theory thus offers a new way to measure behavior as a result of new technology in the form of computer-driven laboratory experiments.

In this chapter, I will present an overview of the concepts of game theory, and in the next chapter, I will present an overview of experimental methods. In each subsequent chapter, I will present and explain the standard game theory concepts and present laboratory experiments that have tested these concepts. I will first define what I mean by the term game theory and then discuss the main behavioral assumption underlying game theory, rationality. Next, I will offer a brief but more precise explanation of what I mean by behavioral game theory. Last, I will describe the different types of games we will encounter in this book.

B. WHAT IS A GAME?

1. GAME THEORY AS AN INTERDISCIPLINARY METHOD

Game theory is a mathematical model of human decision-making. It is an interdisciplinary theoretical method that examines how people (players in a game) make decisions when their actions and fates depend on the actions of other people. Put another way, it examines the strategic interaction of two or more people and how this interaction results in a particular outcome. Game theory is used in the disciplines of economics, political science, sociology, psychology, computer science (artificial intelligence), biology, and mathematics. The appeal of game theory is that it allows researchers to make precise mathematical predictions about institutions (economic, political, biological, family) that are populated by well-behaved players (what I mean by well-behaved will be explained shortly), as well as about the behavioral choices of the players within these institutions.

Game theory is a subset of rational choice theory (or formal theory), which is a catch-phrase for a myriad of modeling traditions that rely on the assumption that human beings act rationally. These various traditions include social choice theory, which examines how the aggregation of preferences results in social outcomes; spatial theory of elections and committees, which examines majority-rule outcomes for situations in which players have spatial preferences (i.e., preferences that are represented in a dimensional space, as will be illustrated in Chapter 3); and decision theory, which examines individual decision-making. While the focus of this book is on game theory, rational choice theories are often intertwined with this area, so I will discuss various elements of these other traditions in this book.

2. GAME THEORY AND EQUILIBRIUM

As in the ultimatum bargaining game discussed previously, a typical game consists of two or more players who have a set of strategies and for whom the application of a particular strategy results in a particular outcome or payoff. Some of the games we will examine will include elements of conflict, whereas others will include elements of cooperation. A game of conflict can be illustrated by most sports games, such as tennis. In this game, one player tries to decide to hit the ball into the left or the right court, and the opposing player tries to anticipate where the ball is going to be hit so that he or she can react to the left or the right. If the first player hits the tennis ball to the left court and the opposing player anticipates right, then the first player wins a point and the other player loses a point. In games with conflict, one player tries to determine what the other player is going to do and selects strategies in order to produce the best outcome for him or herself and the worst outcome for another player. However, not all games involve conflict. Consider the case in which two friends are trying to decide what movie to see. One friend has a preference for an action film, whereas the other friend wants to go to a romantic comedy. Both friends prefer to go to the movie together rather than go to their most preferred movie alone. Hence, this interaction becomes a coordination game in which the friends must cooperate to determine what movie they will see. These two examples show that games can include elements of both conflict and cooperation.[5]

Game theory attempts to solve for solutions to these games of conflict and cooperation by solving for equilibrium given the assumptions of the model (for instance, in the coordination game, a solution may be reached if one friend offers to pay for the movie that she most wants to see). An equilibrium is a solution to a model given the strategic interaction of the players who are within some institutional environment.

An institution or environment is the structured context (often with rules) in which players interact, such as some type of voting system, economic market, legislature, bargaining situation, family unit, or other cultural milieu. I sometimes use these terms interchangeably, but the proper usage is that institutions concern models and environments concern experiments. In essence, however, institutions and environments are both elements that are exogenous to decision-making.

Once an institution is created, behavior can be specified and rules can be set and varied to determine the impact that the variation has on behavior. To determine this impact, we can calculate a solution, or an equilibrium, to the game that is created by the institution. Ordeshook (1986, xiii) observes:

> Game theory is . . . central because it focuses on concepts of equilibrium. In game theory an equilibrium is a prediction, for a prespecified circumstance, about the choices of people and the corresponding outcomes. This prediction generally takes the form "if the institutional context of choice is . . . , and if people's preferences are . . . , then the only choices and outcomes that can endure are . . . " Thus, equilibria replace . . . journalistic interpretation of events.

We can think of an equilibrium in the following manner. Consider a (trivial) voting game in which two players have a choice between three alternatives A, B, and C. Players have preferences such that Player 1 most prefers A and Player 2 most prefers B. It is assumed that players must vote (i.e., there is no abstention), and that they will vote for the alternative that they most prefer. The voting rule is that the players are only allowed to vote for one alternative, and the alternative that does not receive a vote wins. An equilibrium to the game is that Player 1 will vote for A, Player 2 will vote for B, and the outcome will be C. Hence, given the assumptions I have made about the institution and the interaction of the two players,

C is the equilibrium, or the logical and only possible outcome. Game theory is interesting because all games with a finite number of players and strategies have equilibrium solutions. Consequently, it is possible for us to model an almost endless of number of social situations and, given the assumptions we make, find a solution. You may not always like that solution, but it will exist.

3. A GAME IN VON NEUMANN'S SENSE

Although others were actively involved in the development of game theory, the method did not garner academic interest until John von Neumann and Oskar Morgenstern published their seminal *Theory of Games and Economic Behavior* in 1944.[6] As Poundstone (1992) reports, during a taxi ride, a colleague asked von Neumann: "What is game theory?" The narrative of the question and his response are as follows (Poundstone 1992, 6):

> I naturally said to him, since I am an enthusiastic chess player, "You mean, the theory of games is like chess." "No, no," he [von Neumann] said. "Chess is not a game. Chess is a well-defined form of computation. You may not be able to work out the answers, but in theory there must be a solution, a right procedure in any position. Now real games," he said "are not like that at all. Real life is not like that. Real life consist of bluffing, of little tactics of deception, of asking yourself what is the other man going to think I mean to do. And that is what games are about in my theory."

What von Neumann considers a game is illustrated in the following example: Imagine that you and your friend are playing a game in which both of you have two cards, an ace and a king. The object of the game is for your friend to match your two cards, so if you play an ace first and a king second and your friend plays an ace first and a king second, then she matches your cards and wins the game, and you must pay her five dollars. However, your incentive is not to match the cards, so if you think she will play an ace first, you should play a king first, and if you think she will play a king second, you should play an ace first. If you are successful, your friend will owe you five dollars. Say your friend puts two cards facedown so you cannot see them and will not know what card to play. Now you have to place two cards facedown. In what order would you place them? King first and then an ace, or ace first and then a king? The solution to this question is what game theory is all about. There is no computational solution like in chess, but the answer depends on many things, such as what you know about your opponent. Let us say that you know your best friend is obsessed with order, so you believe she will play the ace first, since this is the first card in the deck. But your best friend might think, "My friend knows me well and she knows I have an obsessive personality, so she thinks that since the ace is the first card in the deck I will play this card, but since I know she thinks this, I will play a king and deceive her." However, you reason that your friend will think that you know about her obsessive personality and will try to trick you, so you should play a king first. In this game, what you think (your beliefs) and what you think your opponent thinks (your friend's beliefs) will determine the outcome. In this case, the equilibrium to the game is for each player to randomize the selection and play each card with probability one-half. (The optimal strategy is to play the game blindfolded so that the placement of the two cards is random.) This is the definition of a game in von Neumann's sense.

4. GAME THEORY AND THE IMPORTANCE OF ASSUMPTIONS

The solutions to game theory models are predicated by the assumptions that are posited. The assumptions made in a model are critical because they determine how equilibrium is derived. To illustrate how assumptions can dictate the outcomes of a model, consider the following example about the interdependent nature of driving choices. Imagine you are driving a sports car on a two-lane highway where your speed is eventually slowed by a car in front of you that is pulling a trailer. You have two choices: either follow the car pulling the trailer at a reduced speed or pass the car pulling the trailer. The driver of the car with the trailer has two choices as well: either drive at his current speed or pull over so that you can pass him safely. The outcome of this speed dilemma is a function of the interdependent choices of both players (drivers). We can envision that the worst outcome would be that the car with the trailer would fail to slow down and you would decide to pass it but crash into an oncoming car. One of the best outcomes might be that the car with the trailer would pull over, allowing you to pass. However, both of these outcomes depend on the assumptions that are embedded in the game. If you are late for an employment interview that you cannot miss, then you might risk the dangers of passing the car with the trailer. But what if the game assumes you have a small infant in the car? You would likely not risk the danger of passing the other car. If we assume that the driver of the car with the trailer is in no hurry to get to his destination, then he might be willing to slow down and pull over so you can pass him. But we could also assume that the driver of the car with the trailer also has an important appointment and, as a result of the weight of the trailer, cannot go faster, and slowing down might make him miss his appointment. We can think of a host of other assumptions that would cause each driver to behave differently. The point is that interdependent choices in a game can be influenced by the assumptions specified in the game. In this example, assumptions about risk can determine game play. I will discuss risk in more detail in Chapter 4.

5. RATIONALITY AND SELF-INTEREST IN A CURVED EXAM EXAMPLE

Now let us consider the notion of rationality. One aspect of rationality is that people can be expected to behave in a self-regarding manner, but not necessarily to hurt others. Consider an exam that is based strictly on a curve, for which a student's grade not only depends on how much effort that student puts into studying, but also on how much effort other students put into studying. In this situation, it would be best for a student to exert a high level of effort and study for the exam while other students exert low levels of effort and do not study for the exam. But what is best for *all* students is for everyone to exert low levels of effort and receive a low grade on the exam, curving, for example, a mean of 60 up to a mean of 70. Because of the curve, this result would be similar to the result produced had they all put in a high level of effort and all received a high score on the exam, curving, for example, a mean of 80 down to a mean of 70. Enacting a strategy of low effort is more desirable for all students, since this will free up time to study for other classes (or time to go to a bar!). Knowing that low effort means more free time, the students get together and agree to form a cartel in which they all pledge to exert a low level of effort on the upcoming exam. So the cartel is formed, but the night before the exam one student reasons, "Well, I should study for the exam, because if I do and the others are not studying, then I will receive the highest grade on the exam." However, another student might make the same deduction and study for the exam, and then another, until all the students make the same deduction and all study for the exam, breaking the cartel agreement. Why has the cartel agreement been broken? Because students are motivated to earn higher grades and breaking the cartel gives them the opportunity to achieve a higher grade than the cartel promises. This example illustrates a self-interest motivation on the part of the students, since it is in their self-interest to defect and study when they think other students are cooperating—that is, not studying.

C. BEHAVIORAL ASSUMPTIONS

1. WHAT IS RATIONALITY?

At the heart of game theory are certain assumptions that predict the behavior of players. The standard assumption used has been the concept of rationality, which concerns the preferences players have and how they select among various alternatives.[7] It is important to note that rationality in this context does not concern "real" human behavior, but the behavior we ascribe to the players in our game. We want our players to make humanlike choices, however, so rationality is assumed.

Recall that our games are mathematical models, so players are not people per se but mathematical functions. We need rationality because we are ascribing humanlike behavior to a mathematical function—that is, rationality is needed to make our mathematical functions behave like people. In certain institutions, rationality is an adequate model of human behavior, but in other institutions, other types of behavior predict decisions better.

In this book, I typically refer to two types of people: those who participate in an experiment (subjects) or real human people like me and you. When I refer to a player or players, I am referring to the players we have modeled in our game or the people we have created to make decisions in a game institution. I attribute gender qualities to our mathematical functions by assuming that in a certain game Player 1 is a woman and Player 2 is a man, or Player 3 is a woman, and so on. Because one of the goals of behavioral game theory is to give these mathematical functions more humanlike qualities, we can add emotions as a parameter to the function. For example, what types of emotions (or other behavioral qualities) might differentiate our mathematical functions as being male or female?

The standard model of rationality simply assumes that players have well-defined preferences over alternatives in areas such as food, music, movies, or money. This model also assumes that players derive some payoff, utility, or satisfaction from the various alternatives. Consider the preference for food. Rationality first means that we can assume a player has preferences concerning three different types of foods—say, tacos, hamburgers, or hotdogs. Second, it means that a player is given a preference for each type of food. Third, it means that a player can rank the alternative food choices based on the specified preferences, so that a player might prefer hamburgers to tacos and tacos to hotdogs. Rationality dictates that when a player makes a decision, she will always select the alternative that she most prefers or else she will maximize her utility.[8] It is important to note that because of the interaction of game play, a player may not always be able to achieve the alternative that is highest on her ranked list of alternatives, in which case she will attempt to select the highest available alternative. Let us assume that our player is at a restaurant and orders a hamburger, her highest preferred alternative. She is informed that they are out of hamburgers, so rationality dictates that she will order a taco, her second most preferred alternative, instead. If the restaurant does not have tacos, then rationality dictates that our player will settle for hotdogs. While this selection rule might seem benign, its implementation often leads to outcomes that are self-interested (or self-regarding). In our grading curve example, each student had preferences defined over a particular grade such that an A was preferred to a B, a B preferred to a C, a C preferred to a D, and a D preferred to a F. Hence, when selecting an action, it was in the students' self-interest to defect from the cartel so that they could attain a higher grade as defined by their preferences. Rationality just assumes that a player will select alternatives in accordance with his preferences. In game theory, the assumption of rationality is defined to specify a utility function that converts preferences into a choice. I will discuss utility functions in detail in Chapter 3.

Rationality means that players in the game will make optimal decisions given their preferences and the institution they confront. Rationality dictates how players will behave in a game theory model. No argument is being made that real human behavior fits the standards of rationality in the sense that all choices people make are optimal, because often they are not. Rather, rationality is a modeling technique that assumes that players in a model will make consistent or coherent decisions given the institution in which they are placed.

In a video game there are often non-player characters (NPCs) that are characters in the game not controlled by the user but by the "game artificial intelligence." Noncombative NPCs are often employed in a video game to help move the plot of the story forward, but most NPCs are combative enemies who must be fought and defeated in order to advance in the game. In more advanced video games, these enemies have artificial intelligence (AI) capabilities that, if they are attacked, dictate (through some algorithm) a response such as defending themselves, attacking back, fleeing, or hiding. AI provides only a few algorithms that govern this behavior (or minimum behavior) so that a challenging fight can be had. It wouldn't be a fun game if the enemy did not react to your actions in a manner that you "at a minimum" expected (i.e., if you hurt an enemy, you would expect him to recognize some type of hurt). Furthermore, we would expect all video games that use AI to adopt similar algorithms so that the basic rules governing NPC behavior would be similar and predictable in all games (i.e., in all similar video games, repeated harm inflicted on an enemy would eventually lead to victory).[9] In this way, the user of the game could be familiar with many game environments and would be able to play a game without even reading the instructions, because he would know that all NPCs in all video games would follow some minimum rules of behavior. The reasons for assumptions of rationality that specify minimum rules for behavior are the same.

2. WHY IS RATIONALITY NEEDED?

Rationality is just a set of rules that govern uniform and minimum standards of behavior. This is important for two reasons. First, since the behavior of players is uniform in game theory models, then it is possible to compare results across models. That is, we can determine how comparisons of different institutions impact decision-making, and we can hold human behavior constant when making the comparisons. We can assume that the institutions matter and the people within them do not. However, for certain institutions, the differences in people do matter, and this is why we incorporate psychological variables into the standard model. Second, rationality is a basic framework or baseline for behavior. That is, rationality is the minimum standard of behavior we would expect our players to exhibit. Assumptions of rationality are valuable because they allow us to study other types of behavior. For instance, as noted previously, one of the goals of behavioral game theory is to introduce emotions into players' decision-making calculus. Rationality that is devoid of emotion allows emotions to be studied, because it provides a baseline to study deviations from predicted behavior. Deviations are important because they represent departures from the model that is being studied, and if the model assumes rationality and people are not behaving in a rational manner, then some other motivation (e.g., an emotion not specified in the model) must be the culprit of the deviation.

Clearly, real people do not consistently behave in a rational manner, and mistakes and mishaps are a part of human decision-making. Consequently, rationality allows us to measure players' behavior so that predictions can be made in reference to a particular institutional structure. Recall that a model and its players are all only abstractions of some slice of reality. When players are posited in a model, our concern is only with their choices that are in accord with the preference structure assumed in the model. Consider the baseball stadium example presented at the beginning of this chapter. In the real world, the roof of the baseball stadium would be exposed to different velocities of wind currents over an extended period of time, unlike the short and uniform wind currents to which the model stadium was exposed in the test. Nevertheless, the simulated wind currents provided an adequate test of the model that allowed useful predictions to be made. We can think of the assumption of rationality like the simulated wind in the baseball stadium example: it is not the real thing, but it is a close approximation. It is a reasonable theory of people who have only one incentive (a lust for more), and this motivation provides a nice framework to explore different theories of behavior. In the baseball stadium example, we can imagine the researchers saying, "Now that we have our model established, let's see what other factors impact the flight of the ball, such as the noise level or heat generated from fans." If this were the case, then our researchers would have to create proxies for people or fans in their model.

The big question is: What is the value of assuming models with mathematical players and then trying to make inferences for human decision-making behavior? It is argued that since assumptions of rationality do not model the complexity of real human behavior, then no inferences can be made about real human behavior at all. But the point is that real human behavior is too complex to study, so we need to simplify it in order to examine the various elements that are involved and how they work. Whether or not the inferences that are drawn from our mathematical players are valid is an empirical question.

Experiments allow us to examine how the behavior of people in a model conforms with or differs from real human behavior. Instead of taking my word for it, I hope that the material in this book will show how game theory and experiments have contributed a wealth of knowledge about human decision-making. If you read the entire book, then I will let you be the judge of whether we have learned anything about real human behavior from these abstract models.

D. BEHAVIORAL GAME THEORY

1. RESEARCH METHODS OF BEHAVIORAL GAME THEORY

This book takes a behavioral game theory approach that attempts to add more human qualities to the basic rational player that is assumed. Colin Camerer, the scholar who coined the term behavioral game theory and wrote an influential book by the same name that effectively made game theory a distinct discipline within the field of social sciences, comments (2003, 3):

> Behavioral game theory is about what players *actually* do. It expands analytical theory by adding emotion, mistakes, limited foresight, doubts about how smart others are, and learning to analytical game theory.... Behavioral game theory is one branch of behavioral economics, an approach to economics which uses psychological regularity to suggest ways to weaken rationality assumptions and extend theory.

Innocenti and Sbriglia (2008, 1–2) further comment:

> Game theorists would have no difficulties admitting that there is overwhelming evidence that human beings, more often than not, do not behave as predicted by the theory. This happens because an individual's decision—when involved in strategic interactions—is affected by their cultural and historical background, their feelings and psychological attitudes and their ethical values. In other words, human players play games in a human way, and the final outcome is likely to be affected.

These two quotes illustrate that behavioral game theory is concerned with how real people make decisions when confronting various institutional mechanisms assumed in game theory models. Behavioral game theory attempts to understand how psychological factors such as emotions can predict decision-making behavior in different institutions (see Camerer 2003). It is important to note that most behavioral game theorists are not attempting to discredit game theory by showing real people's departure from the theory; rather, they are attempting to build a better theory of human decision-making using game theory as a method.

Behavioral game theorists use an observational approach to their research method. They observe outcomes of an experimental test of a model and study subjects' actual behavior before revising their theory based on whether that behavior conforms with or deviates from the model's prediction. For instance, in the ultimatum bargaining example, we saw that subjects could exhibit behavior that resembled fairness although it ran contrary to the assumptions of rationality. We can imagine a researcher observing an experiment in which some subjects' behavior corresponds to the predictions of rationality while others' behavior appears to be a result of emotions such as fairness. This researcher might conclude that because subjects exhibit behaviors corresponding to ideas of rationality and fairness, the utility function or preferences of these players should probably include a term for rationality and a term for fairness. Behavioral game theory, as will be discussed in Chapter 3, introduces notions such as social preferences to model emotions besides rationality within a game theory model.

This observational research method of observing experimental outcomes and theorizing is the research method used by behavioral game theorists. The general research process is as follows:

1. Consider a standard game theory model (that assumes rationality)
2. Conduct an experiment on the standard game
3. Observe outcomes of the experiment
4. Note any outcome that deviates from the prediction of the standard game
5. Theorize about the cause of the deviation and construct a revised model based on the conjectures about the cause of the deviation
6. Conduct a new experiment on the revised model
7. Observe outcomes of the experiment to determine if the deviation is explained (if not, repeat starting at Step 5)[10]

This process illustrates why experiments constitute the methods side of behavioral game theory, since they allow for this testing, observing, and remodeling process. An experiment allows a researcher to specify the assumptions of a model in a controlled environment and populate that model with real people or subjects, allowing these real people's behavior in the game environment to be measured. Although some incentive constraints are placed on the subjects' behavior, players are free to interact within the environment and play the specified game to determine if the model's equilibrium predictions are realized. If not, then the deviations in behavior from the rational choice model can be studied. Again, any deviations from the equilibrium (or correct) behavior could be the result of many factors, including psychological ones such as issues related to fairness.

2. HISTORICAL DEVELOPMENTS IN BEHAVIORAL GAME THEORY

Behavioral game theory deals with irregularities that occur when assuming rationality. The earliest examples of irregularities in the theory are the Bernoulli or St. Petersburg's Paradox (Bernoulli 1738), which I will discuss in Chapter 4, and the Allais Paradox (1953). These paradoxes showed that certain assumptions of rationality could be rendered inconsistent. Kenneth May (1954) presented an experiment that showed that the ordering of alternatives can contradict rationality assumptions (we will examine this experiment in Chapter 3); Slovic and Lichtenstein (1983) later presented a more rigorous test of these findings. The psychology and economics duo of Amos Tversky and Daniel Kahneman (1986a and b) also showed how rational choice decision-making can be skewed by framing effects (we will also discuss this contribution in Chapter 4), earning Kahneman a Nobel Prize in 2002. Charles Plott (1976) showed inconsistencies in rational choice assumptions when examining certain voting rules.[11]

An important experiment by Güth, Schmittberger, and Schwarze (1982) first showed that when a subject's identity in an ultimatum bargaining game was anonymous, unfair offers were often rejected—contrary to the model's prediction. These researchers' results challenged the assumption that people in these types of bargaining situations are self-regarding. The experimental results showed that people do have preferences for money, but they also have preferences for other things that are not controlled for in the experiment, such as fairness. This finding prompted a barrage of experiments testing their results (some of which are presented in Chapters 4 and 8). Subsequent experiments established the key idea that in certain game environments, people have social utility functions (or payoff-interdependent preferences).

In short, behavioral game theory focuses on how people actually play theoretical games and adds parameters to our mathematical functions to make behavior more humanlike. It is interested in issues related to framing, or how the environment of a decision can bias a decision (discussed in Chapter 4). It is about the feelings and emotions that are present during a decision, and about how to model thinking and learning in terms of decision-making (discussed in Chapter 11).

Summing up the goal of behavioral game theory, Camerer (2003, 465) succinctly states:

> Two major criticisms of game theory: first, that game theory assumes more calculation, foresight, perceived rationality of others, and (in empirical applications) self-interest than most people are naturally capable of; and, second, that in most applied domains there is too much theorizing about how rational people *would* interact strategically, relative to the modest amount of empirical evidence on how they *do* interact....
>
> Both criticisms can be addressed by observing how people behave in experiments in which their information and incentives are carefully controlled. These experiments test how accurately game-theoretic principles predict behavior. When principles are not accurate, the results of the experiment usually suggest alternative principles. This dialogue between theory and observation creates an approach called "behavioral game theory," which is a formal modification of rational game theory aided by experimental evidence and psychological intuition.

E. DIFFERENT TYPES OF GAMES

There are many differences in the types of models studied using game theory. In this introduction, I would like to discuss some of these differences, with the understanding that this list includes the games covered in this book but is not comprehensive.

1. COOPERATIVE VS. NONCOOPERATIVE GAMES

First, game theory can be classified into two broad classes of games: cooperative games and noncooperative games. In cooperative games, players can communicate with each other and form binding coalitions and pacts, or agreements among members to coordinate any strategic action. These games are based on how players will divide up aggregate payoffs. In noncooperative games, players cannot form binding agreements, and they may or may not be able to communicate with each other. Some scholars have argued that noncooperative games are more fundamental than cooperative games, because cooperative games, unlike noncooperative games, assume binding agreements, which are difficult provisions to enforce in the real world. Commenting on the differences between cooperative and noncooperative games, Morrow (1994, 76) notes:

> One of the great disputes in game theory concerns the primacy of noncooperative games. Harsanyi and Selten (1988) contend that noncooperative games are more fundamental than cooperative games because they explicitly model the means of enforcement of agreement. Cooperative games, they argue, make it too easy for the players to make agreements; the players can bind themselves to agreements that may not be enforceable. Noncooperative games force us to consider how collaboration among players is implemented in the game and what incentives the players have to violate such agreements.

As a result of this argument and the fact that most models studied today are noncooperative games, this book will focus exclusively on noncooperative game theory. For a review of cooperative games, see Ordeshook (1986).

2. COMPETITIVE VS. NONCOMPETITIVE GAMES

Games can also be classified into strictly competitive games and games that are not strictly competitive. In a strictly competitive game with two players, one player's payoff decreases when another player's payoff increases. A game in which the payoffs to players sum to zero is referred to as a zero-sum game. For instance, a competition between two candidates for a political office is a strictly zero-sum game, because one candidate must win and the other must lose. However, the payoffs in strictly competitive games do not always sum to zero, and when they do not, the game is referred to as a constant sum game. Hence, zero-sum games are just a subset of constant sum games. It is important to note that in strictly competitive games, there is no opportunity for compromise or joint gains. But many social situations are not strictly competitive and often involve coordination and cooperation. These games are referred to as variable sum games (or non-zero-sum games), or games in which the payoffs to the players do not sum to zero. In other words, one player's defeat is not always another player's victory, as in the movie date example mentioned earlier. I would like to emphasize here an important and often confusing point: just because one player wins six dollars and another player wins four dollars does not mean that the player who gets only four dollars loses the game and the player who gets six dollars wins the game. This outcome only means that this distribution is the best outcome that a player could achieve given the game theoretic environment in which he or she is placed. Chapters 5 through 7 will examine both constant sum and variable sum games in greater depth.

3. NORMAL FORM VS. EXTENSIVE FORM GAMES

There is also a distinction between normal form and extensive form games. Normal form games are essentially games presented in tabular form (e.g., 2 × 2, 3 × 2 rows and columns), whereas extensive form games are presented in a game tree, which depicts moves as progressing from one branch to the next. The difference in formulation allows game theorists to consider different decision-making environments. Normal form games generally model situations in which decisions are made simultaneously and players do not know what strategies other players have selected when they select their own strategy. Extensive form games allow for more dynamic games in which a player knows how another player has moved before he or she makes a strategy decision. Extensive form games can also contain elements of simultaneous choice; this will be clarified in later chapters. Chapters 5 through 7 will consider normal form games, and Chapters 8 through 10 will cover extensive form games.

4. PURE VS. MIXED STRATEGY GAMES

A distinction can also be drawn between pure strategy and mixed strategy games. In some games, equilibrium is achieved when a player selects a pure—or only one—strategy. For example, if a player has a choice between strategy X and Y, he or she must select either X or Y, and not both. This is in contrast to games that solve for equilibrium through a player's mixed strategy. We can think of mixed strategies as a situation in which a player selects strategies but the outcomes are determined by lottery, or as a probability distribution defined over outcomes. In this sense, instead of a player choosing just X or Y, he or she will select X with a certain probability p (where $p \geq 0 \geq 1$) of achieving some outcome, or Y with a certain probability $(1 - p)$ of achieving some outcome. For example, a player might select X 40 percent of the time and Y 60 percent of the time in order to win a prize z. Games that solve for equilibrium using pure strategies will be covered in Chapters 5 and 6, and games that solve for equilibrium using mixed strategies will be covered in Chapter 7.

5. SINGLE-SHOT VS. REPEATED GAMES

There is also a distinction between single-shot games and repeated games. As the term suggests, in a single-shot game, players only play the game once. These games are usually simultaneous move games in which the players do not know what strategies other players have selected before they have to make their own selections. Repeated games can be of two types: finite repeated games, in which an ending period is fixed (e.g., the game lasts for only 10 periods), or infinite repeated games, in which no ending period is fixed (e.g., there could be some probability dictating when the game will end). Repeated games involve a time element, since the game is played more than once. A stage refers to a single period of a repeated game (the terms stages, periods, and rounds are often used interchangeably). A stage game refers to a game that is played in each period. This book will mostly be concerned with single-shot games, but I will discuss finite repeated games in Chapter 6.[12]

6. COMPLETE AND PERFECT INFORMATION VS. INCOMPLETE AND IMPERFECT INFORMATION

Games can be further divided based on the information that players have about the game structure. First, players can have knowledge about all aspects of the game. Two informational structures are used to model this knowledge: complete information and perfect information. Complete information (often referred to as full information) is a situation in which players know the strategies and payoffs of other players. This condition is usually assumed in simultaneous move games. Perfect information is a situation in which a player knows the moves (actions) of other players, as well as the complete history of moves. Perfect information is usually assumed in extensive form games. Some sequential move games involve both complete and perfect information, meaning that everything about the game structure (strategies, payoffs, and moves) is known to all players. Second, in a situation of imperfect and incomplete information, players are not privy to information about all relevant aspects of the game. Imperfect information means that a player does not know all the moves or actions in a game, and incomplete information means that players lack knowledge about the strategies and payoffs of other players. In games of both imperfect and incomplete information, players do not know the strategies and payoffs of other players, nor do they know their moves. Games can also assume complete and imperfect information when strategies and payoffs are known but not the moves of another player, or incomplete but perfect information when strategies and payoffs are not known but moves are. Chapters 5 through 9 will be concerned with complete and perfect information games, and I will introduce games of imperfect and incomplete information in Chapter 10.

F. SUMMARY

This chapter has provided an overview of the basic concepts of game theory. In this context, games model the strategic interaction of two or more players, and a game's equilibrium or outcome is determined by the strategies players select and their beliefs about the game environment. These games differ from games like chess, which is a computational game. Rationality is an assumption common to all rational choice models, of which game theory is a subset. Rationality means that players in the game will select among alternatives that best benefit themselves. Consequently, this behavior can result in a negative outcome for another player. This book takes a behavioral approach to game theory, meaning that I will examine laboratory experiments that have empirically tested the various game theory concepts. Behavioral game theory attempts to add a psychological explanation to the deviations found in experiments on standard game theory models. As we will see in future chapters, experimental methods are an ideal way to test notions of game theory, because they are able to replicate the parameters of the model in an experimental setting and subsequently allow real people to populate the model and play the game to test its consequences. In the next chapter, I will provide an overview of experimental methods that will help us to understand the various experiments that are presented in this book.

NOTES

1. This story is recounted in Associated Press (1991).
2. Empirical verification means that a model has been tested by some sort of observational or non-observational data. This term assumes a scientific approach to testing the validity of models.
3. Maybe you could randomly alternate the two grips 10 times each among the 20 strangers, but would this be any different from meeting the strangers randomly and gripping the first 10 tightly and the second 10 gently?
4. The question we must ask is: Are bad data better than no data?
5. I will discuss this distinction further in Section E of this chapter.
6. I provide a brief history of game theory in Appendix 2.

7. There are several different definitions of rationality, such as procedural rationality, which is the process of making decisions. In this book I use the traditional definition, which is often termed substantive rationality (see Schwartz 1998 for a discussion).

8. This assumes that a person is making a decision at a particular point in time t, which could be different at another point in time t + 1. That is, a person might prefer hotdogs in the afternoon but tacos in the evening.

9. See Chapter 11 for a further discussion of video games in regard to beliefs based on conditional probabilities.

10. Steps 1 through 3 typically involve using another researcher's published results.

11. Also see Abrams (1980) for a host of examples on inconsistencies.

12. For a discussion of infinite repeated games, see Fudenberg and Tirole (1991) and Morrow (1994).

CHAPTER 2

WHAT ARE LABORATORY EXPERIMENTS?

A. WHY EXPERIMENTS?

1. BEN FRANKLIN'S CLOTHES EXPERIMENT

What is an experiment? In a basic sense, an experiment is a way to test an idea or question you might have. Consider a question posed by Benjamin Franklin: Given the weather conditions outside, what are the most comfortable clothes to wear in terms of color? What is a reasonable way to empirically answer this question? First you have to obtain a measure for comfort levels of different colors of clothing in different weather environments, and then you could observe how the comfort of different types of clothing varies with the weather conditions. However, this approaches requires a measure for comfort. One way to obtain this measure is to wear different colors of clothing in different types of weather conditions and subjectively determine and record which clothes are more comfortable given a particular weather condition. But this procedure would be very difficult to implement, and the results would be biased, because the measure of comfort is subjective.

In responding to this question, Benjamin Franklin instead used an experimental approach to understand the relationship between the color of clothing and weather conditions. His measure of comfort was derived from an objective evaluation of the rate at which the sun penetrated different colors of clothing (light vs. dark) to produce different heat levels. This difference in heat levels could then be used to build a measure for comfort. Franklin reasoned that if the sun penetrated clothing too much, it would be too hot and not very comfortable, and if the sun did not penetrate clothing enough, it would be too cold and also not very comfortable.

Franklin went to a tailor's shop and obtained a cloth book that had different colors of cloth samples. Tailors use these books to show options to customers who are deciding on colors and types of fabric for clothes they want to be made. Franklin separated out the different colors of cloth and laid them all on top of the snow on a sunny day for several hours. When he returned to his "experiment," he noticed that the dark colors had sunk further into the snow than the light colors because the heat had penetrated the dark clothes more than the light ones. Franklin thus concluded that dark clothes should be worn in the winter since they absorb more sunlight, and light clothes should be worn in hotter weather since they resist the sun more (Van Doren 1945).

While we can debate the merits of this experiment (Franklin disregards the thickness of the samples, for example), the point I want to make here is that empirical questions that are not accompanied by empirical observations remain just questions. Benjamin Franklin's experiment provided empirical observations, and we can probably conclude that his findings are accurate, given the fashion trends of today that highlight dark clothes during cold weather and light clothes during hotter seasons.

In game theory, empirical observation played only a small role until recently, when political economy experiments began to test the various game theory models. The term political economy experiments is a misnomer, because these tests have been conducted not only by economists and political scientists, but also by psychologists, sociologists, biologists, and neuroscientists seeking to test rational choice models.[1]

2. THE NEED FOR EXPERIMENTS AND THE GROWTH OF EXPERIMENTS

Although we can document political economy experiments dating back to the 1930s, there was not a systematic effort to conduct experiments on game theory concepts until the 1980s.[2] What changed? I would argue that three things changed. First, starting in the 1970s, there was an explosion of formal theoretical results produced by rational choice theorists. These models offered predictions of outcomes, but the results often conflicted with other similar models, or models contained multiple equilibria. The question was how to determine which models were more accurate or which equilibria were more prevalent. When a researcher needed empirical verification to test the predictions of a model, observational or field data were often not available. I will define what I mean by observational or field data more precisely later in this chapter, but for now, just think of these data as data that are naturally generated from real-world observations. For example, statistics on the unemployment rate are observational data, and while market forces do influence this rate, the rate itself occurs naturally as a result of the number of people who are out of work. Hence, a researcher does not influence this rate, but only observes it. Because some game theory models were so precise in their specifications, no naturally occurring real-world situations could replicate the model's parameters. Consequently, no observational data were readily available to allow researchers to test those models that required empirical validation. Camerer (2003, 20–21) notes:

> Sherlock Holmes said, "Data, data! I cannot make bricks without clay." Experimental results are clay for behavioral game theory. The goal is not to "*dis*prove" game theory (a common reaction of psychologists and sociologists) but to improve it by establishing regularity, which inspires new theory. Without some sort of observation, theoretical assumptions are grounded in causal pseudo-empirical work—informed opinion polls in seminar and office discussion and using one's intuition (a one-response poll). Biologists don't just ask "If I was a robin foraging for food, how might I do it?" They watch robins forage, or ask someone who has.

What Camerer calls for is precisely what was produced: empirical verification of the assumptions and predictions of rational choice models using laboratory experiments. Since field data or observational data can at best provide a rough estimation of a model's prediction, then the only recourse was to conduct experiments designed according to the specifications of the model, as Camerer suggests. Plott (1990, 8–9), also commenting on the need for experiments in the absence of observation or field data, notes:

> If the data can only come from the field, and from repeated observations of special circumstances found there, then the appropriate tests will probably never be conducted. The only practical source of data that can be obtained within an appropriate time frame and serve as a guide for many of the newly developed theories is the laboratory.

Hence, the 1970s saw both an increase in game theory models and an increase in the use of laboratory experiments to test the predictions of these models. Reflecting on the important contribution that experiments can make, Palfrey (2006, 915) states:

> Researchers who were trained primarily as theorists—but interested in learning whether the theories were reliable—turned to laboratory experiments to test their theories, because they felt adequate field data were unavailable. These experiments had three key features. First, they required the construction of isolated (laboratory) environments that operated under specific, highly controlled, well-defined institutional rules. Second, incentives were created for the participants in these environments in a way that matched incentives that existed for the imaginary agents in theoretical models. Third, the theoretical models to be studied had precise *context-free* implications about behavior in any such environment so defined, and these predictions were quantifiable and therefore directly testable in the laboratory.

A second explanation for the increase in the use of experiments to test models is that experimental studies beget more experimental studies. Often the results of an experiment will produce more questions than answers, prompting a researcher to conduct more experiments. Other times, several researchers will have different views on how a theory should be tested, leading many different experiments to be conducted on the same theory. In some cases, an experiment will only test one aspect of a theory, leaving other aspects open to experimentation. Concerning the emergence of the laboratory method in economics, Plott (1990, 14–15, 918) prophetically observes:

> The famous law of J. B. Say, that supply creates its own demand, seems applicable to the case of experimental research. The application of experimental methods generates research questions that can only be answered by a more intense use of experimental methods. The supply of experimental research creates a demand for even more experimental research. When Say's law of experimental methods takes over, the stage is set for an ever increasing tendency to use experiments. The stage becomes set for economics to slowly but surely become a laboratory experimental science....
>
> To those who are not enthusiastic about the use of laboratory experimental methods, the prohibitionists so to speak, the good news is that the profession has tasted the devil's brew, the use of experimental methods, and likes it....Those who have not

been touched are being tempted. Say's law of experimental methods seems to be operating everywhere. The impact might not be noticeable yet but the process is operating.[3]

As Plott points out, the results of one experiment inevitably call for another experiment to be conducted, and then another, and so on. This process has continued, so that now we have a framework based on standard models on which we can build new and more interesting models. In effect, behavioral game theory is a second wave of experimental methods that is itself a result of this initial explosion of political economy experiments, as predicted by Say's law of experimental methods.

A third reason for the increased use of experiments was technological advancement primarily related to computers. Early experiments were called pen-and-paper experiments because subjects would write their responses to stimuli on a piece of paper and hand it to an experimenter. This process was time-consuming, because the experimenter would then have to record the responses and maybe make some calculations before the process could repeat itself. While some experiments are still conducted in this manner, the majority (at least of political economy experiments) are now conducted in a computer laboratory where a wireless or hardwired network connects the machines. Subjects respond to a computer interface that displays experimental stimuli, records responses, and so on. New computer technology has allowed for greater flexibility in design and an increase in the number of experiments that can be conducted in a single experimental session. Researchers with little programming experience can even find free, canned software programs to create a network-based experiment.[4] Laptop computers or smart phones also allow researchers to conduct experiments in remote villages in third-world countries.

As a result of these three developments, we have seen an explosion of experiments that study game theory models. One of the purposes of this book is to expose you to this important shift. In this chapter, I will provide a broad overview of experimental methods. First, I will define what an experiment is. Next, I will discuss the purpose of experiments—establishing causality. Third, I will address issues of experimental validity. Finally, I will describe the different types of methods that are used to conduct experiments.

B. DEFINING A LABORATORY EXPERIMENT

1. WHAT IS A LABORATORY AND HOW DOES IT DIFFER FROM THE FIELD?

A laboratory is a rather simple term to define. In the classical definition, a laboratory is the central location at which an experiment is conducted, where subjects participate in the experiment. Computers may be involved, but not always. The laboratory might have more than one room that allows the experimenter to separate subjects or, in the case of a computer laboratory, partitions to prevent subjects from viewing each other's screens. For experiments conducted on a university campus, the laboratory is essentially a single location on campus. As a result of technological advances, however, the idea of all experimental subjects congregating at a single location is dated. Today, laboratory experiments are not restricted to a single location, and lab-in-field experiments (defined in the following paragraphs) can be conducted anywhere in the world.

Laboratory experiments differ from field experiments, which, as the name suggests, are conducted somewhere other than a room on campus. In field experiments, a researcher attempts to control and manipulate some aspect of the naturally occurring environment and observe subject behavior in response to that manipulation. For instance, a researcher in a voting experiment might give one set of voters some voting literature and give nothing to another set of voters. The researcher could then determine if the individuals who were given the literature were more likely to vote compared with the individuals who received no literature. In this field experiment, the researcher would be observing a real election, with his only manipulation or control being the distribution of the voting literature (see Gerber and Green 2000). In contrast, laboratory experiments attempt to control all possible aspects of the experimental environment and then observe behavior when aspects of that environment are manipulated. Lab-in-field experiments are a hybrid of the two types (Morton and Williams 2010). These experiments are conducted outside of a computer laboratory, but portable electronic devices (such as handhelds or laptop computers) are used to maintain some of the controls found in the laboratory.

2. WHAT IS THE DEFINITION OF AN EXPERIMENT?

The definition of an experiment is not so simple. Although we probably all have a notion of what an experiment is, different scholars have different definitions. In the traditional definition, an experiment is a study that establishes controls over and deliberately intervenes in the question of interest and involves the random assignment of subjects to different treatments, with a treatment being some stimulus given to subjects (Cook and Campbell 1979; Shadish, Cook, and Campbell 2002). According to this definition, a researcher must first randomly assign subjects to different treatments (I will explain why this is important in the next section). Next, one group of subjects is exposed to Treatment A and another group is exposed to Treatment B so that analysis can compare the behavior of subjects in the different groups. For example, one group of subjects might be assigned to read a Scooby Doo comic book and another group of subjects might be assigned to read passages of *Foucault's Pendulum* by Umberto Eco before both groups are assigned to complete a math quiz. A researcher can then compare the two groups to determine whether the subjects who read passages from *Foucault's Pendulum* did better on the quiz than subjects who read the Scooby Doo comic book. A baseline group is usually formed to which the researcher can draw comparisons; this group does not read passages from either a comic book or *Foucault's Pendulum*, but simply takes the math quiz. This baseline allows treatment variables to be compared.

In order to measure the effect of the treatment, this method compares subjects who have not been treated with subjects who have been treated to determine the impact of the treatment on behavior. This knowledge is obtained by subtracting the mean behavior of the untreated subjects from the mean behavior of the treated subjects. If the difference is near zero, then the treatment has no impact, whereas a large difference in mean behavior implies a larger impact of the treatment. For example, assume that an IQ test is administered to all subjects, and subjects who read the comic had a mean IQ of 100, subjects who read Eco's book had a mean IQ of 106, and subjects who read neither had a mean IQ of 103. Hence, the treatment effect on subjects who read Eco's book would be a positive 3, and its effect on subjects who read the comic would be a negative 3. You could then conclude (given the appropriate statistical tests) that reading Eco's book has a positive impact on IQ while reading a comic book has a negative impact.[5]

While I agree that this procedure describes an experiment, the standard definition of an experiment is narrow and excludes many experiments used to test various game theory concepts, as well as Franklin's experiment.[6] For instance, two players playing a prisoner's dilemma game in a laboratory would not be considered an experiment, since they are not exposed to different treatments or a baseline.

Morton and Williams (2010) adopt a more general definition of an experiment. To explain their definition, I must first explain how research is conducted in nature, or using what is called observational data. The process of generating data is referred to as the data-generating process (DGP) and is used in collecting census data, crime statistics, historical archival data, survey data, or any other type of data that a researcher analyzes. When observational data are collected, the researcher only observes the data being generated and does not intervene in the process by which they are generated. For example, if the researcher is interested in census data, which provides information about the socioeconomic makeup of households, he or she does not try to intervene in the process of how the data are generated by rearranging the composition of households but only waits patiently until the results are compiled, after which the researcher analyzes the data. Researchers do not attempt to manipulate the DGP. In contrast, researchers who use experimental data attempt to manipulate the DGP. Instead of patiently waiting for data to be created, the researcher actively and deliberately attempts to control how the data are generated by establishing controls in the laboratory.

The definition of an experiment used in this book assumes that a researcher has deliberately intervened in the DGP. Morton and Williams (2010, 42) note:

> In an experiment, the researcher intervenes in the DGP by purposely manipulating elements of the environment. A researcher engages in manipulations when he or she varies parts of the DGP so that these parts are no longer naturally occurring (i.e., they are set by the experimenter). We might imagine an experimenter manipulating two chemicals to create a new one that would not naturally occur to investigate what the new chemical might be like. In a laboratory election experiment with two candidates, a researcher might manipulate the information voters have about the candidate to determine how these factors affect their voting decisions. In both cases, instead of nature choosing these values, the experimenter chooses them.

According to this definition, then, two players playing a prisoner's dilemma game in a laboratory would be considered an experiment, as would Benjamin Franklin's test. The more important question is why a researcher would want to manipulate the DGP. The answer is simple: to establish causality. I will discuss this response in the next section.

C. ESTABLISHING CAUSALITY

1. RANDOMIZATION OF SUBJECTS TO TREATMENTS AND EXPERIMENTAL CONTROLS

Causality is the impact that one variable has on another variable or variables, where variables are concepts that can be measured. For example, the previous IQ example demonstrates that the type of material read before taking an IQ test can impact behavior. This is a causal relation. However, the presence of a causal relationship simply means that most of the time we will see this relationship, but once in a while a person who reads the comic book will score higher on the IQ test than a person who reads Eco's book. In the case of the IQ test, we do not know if reading certain books and scoring high on the test is a valid causal relationship, because I have not described the process by which the measures (i.e., the IQ numbers) were obtained.

In laboratory experiments, the establishment of causality is a two-step process: first, subjects are randomized to treatments, and second, controls are established over the experimental environment (I will discuss each step in detail in the following sections).[7] If these two elements are perfectly controlled for and we can establish a relationship between two variables, then we can be confident that this relationship is true and unaffected by exogenous factors that could bias its measurement. But these two elements are not always perfectly controlled for, so researchers have to decide the best way to implement the desired controls given their environment. Sometimes, as you will see in later experiments, the lack of controls is the most interesting part of the experiment. No claim has been made that experimental methods perfectly measure causality, because they do not; rather, they are the clearest empirical test of game theory models that we currently have, given the technology that we have at our disposal. Hence, if there is value in these types of models, then there has to be the same value in this method.

The first part of causality is established by randomizing subjects to different treatments and comparing different treatment variables to determine how they differ when all other parameters of the experiment have been controlled for. Why do we need to randomize subjects to different treatments in order to establish causality? The idea of randomizing subjects to different treatments is to control for the innate characteristics of subjects (e.g., political preferences, educational levels, income levels, age, race, gender). If we can randomize subjects who have these characteristics into different treatments, then subjects in the different treatments will have the same innate characteristics, which will allow the researcher to make comparisons across all groups (assuming innate characteristics are the same or controlled for).

The second part of causality is establishing and maintaining controls over the experimental environment. This step is very important, because instituting controls over the experimental environment is how a model is replicated within the laboratory. Morton and Williams (2010, 44) state:

> A researcher engages in control when he or she fixes or holds elements of the DGP constant as he or she conducts the experiment. A chemist uses control to eliminate things that might interfere with his or her manipulation of chemicals. In a laboratory election, if a researcher is manipulating the information voters have about candidates, the researcher may want to hold constant how voters receive information and how much of the information voters have so that the researcher can focus on the effects of information on how voters choose in the election.

Establishing controls mean controlling confounding factors, or factors that can hinder causal statements. Controlling these factors allows us to observe the relevant part of the model that is being tested. In the next sections, I will examine randomization and discuss experimental controls in more detail. Again, in laboratory experiments, both of these components are needed to establish causality.

2. EXAMPLE OF THE IMPORTANCE OF RANDOMIZATION OF SUBJECTS TO TREATMENTS

Imagine that we have three types of subjects. Say that subjects who are incredibly rich are assigned to one treatment, subjects who have average incomes are assigned to another treatment, and subjects who are incredibly poor are assigned to a third treatment. Would a researcher be able to make comparisons about outcomes across the three treatment groups? The answer would be no, since incredibly rich subjects, average income subjects, and incredibly poor subjects would probably behave differently. To explore this point further, let us revisit the ultimatum bargaining game we discussed in the first chapter.

Recall that in this game, two players bargain over a finite resource such as 10 dollars, and one player makes an offer to another player over the allocation of the 10 dollars. If the second player rejects the offer, then both players receive nothing. The equilibrium prediction of this game is that the first player should offer an allocation that benefits himself to the detriment of the second player. In laboratory experiments, various treatments could be varied to determine under what conditions the first player would make fair offers. A researcher might conjecture that different types of communication among the bargainers would influence whether fair offers were made. The researcher could design a three-treatment experiment in which Treatment 1 allows subjects to meet and communicate verbally, Treatment 2 allows subjects to communicate with each other via text messages, and Treatment 3 does not allow subjects to communicate at all.[8] The dependent variable, or the variable the researcher is attempting to explain, would be whether a subject sent a fair or unfair offer. Hence, the type of communication allowed would be the independent variable, or the variable the researcher is using to explain variations in the dependent variable.

Now assume that our three subject groups are assigned to one of the three treatments. Say that the incredibly rich group is assigned to Treatment 1, the average income group to Treatment 2, and the incredibly poor group to Treatment 3. What if the results show that in Treatment 1 ultra-fair offers are made all of the time (i.e., the first player offers almost all of the 10 dollars to the second player), in Treatment 2 fair offers are made half of the time, and in Treatment 3 no fair offers are made. Could we conclude that more communication allows fairer offers? No, because it is impossible to rule out the possibility that regardless of the treatment to which the incredibly rich group is assigned, they will make ultra-fair offers all the time, because 10 dollars is just a trivial sum to them. Similarly, we cannot rule out the possibility that regardless of the treatment to which the incredibly poor subjects are assigned, they will always make unfair offers, since 10 dollars is a substantial amount of money for them. Hence, it is impossible to determine the effect of communication on the rate of fair offers because of the income bias inherent in these groups. However, randomization of subjects in the three treatments would eliminate this bias, since incredibly rich, average income, and incredibly poor subjects would be mixed with each other, canceling out the effects of income levels and removing them as a factor in the analysis. If randomization of subjects across treatments is done correctly, then the impact of innate characteristics can be ruled out of a causal relationship.

3. EXPERIMENTAL CONTROLS AND CONFOUNDING FACTORS

The second step in establishing causality is considering confounding factors, or factors that may interfere with an experimental treatment (or manipulation).[9] Controlling these factors allows us to build models in the laboratory. For example, looking at the no-communication treatment in the previous bargaining experiment, a confounding factor could be unanticipated eye contact between two subjects that reveals information regarding their choices to each other. This act would effectively turn the no-communication treatment into a treatment that allowed communication through sneaky signals. If we can control these and similar factors, then we can say that we have reconstructed our communication model (or at least the aspect in which we are interested) in the experimental environment. Morton and Williams (2010, 43–44) comment:

> Experimenters worry (or should worry) about factors that might interfere with their manipulations. For example, trace amounts of other chemicals, dust, or bacteria might interfere with a chemist's experiment. That is, the chemist may plan on adding together two chemicals, but when a trace amount of a third chemical is present, his or her manipulation is not what he or she thinks it is. . . . A chemist uses control to eliminate things that might interfere with his or her manipulation of chemicals.

An experiment is conducted to control confounding factors so that a relationship between variables can be more clearly established. To illustrate what I am talking about, consider a modified version of the matching game that we discussed in the first chapter. In this game, Player 1 and Player 2 both have an ace and a king. Player 1 selects one of her cards and lays it facedown, as does Player 2. If the cards match, then each player wins two dollars, and if the cards do not match, then neither wins anything. In this game, before Player 2 lays down his card, Player 1 has the option to send him a nonbinding message indicating what card she plans to lay down ("nonbinding" means that Player 2 is not committed to play what her message indicates she will play). In order to send a message, Player 1 has to pay 50 cents. If Player 1 does not send a message and both players' cards match, then each player will get two dollars. If Player 1 pays for a message and both cards match, then Player 1 gets $1.50 (two dollars minus the 50-cent information cost) and Player 2 gets two dollars. Whether or not Player 1 buys a message depends on her notion of the fairness of payoffs. First, notice that regardless of whether Player 1 buys information or not, if the cards do not match, then neither player receives anything. Consequently, there is no risk involved in this outcome, since the information cost of 50 cents is erased if the cards do not match. The structure of this game gives Player 1 an incentive to buy information and send a message if she does not care about Player 2's payoff, because regardless of that payoff, she will net $1.50. But if she thinks that a situation in which she alone bears the cost of the message and payoffs are unequal is unfair, then she might not purchase the information and might rely instead on a random outcome. In this game, then, the action of purchasing information reveals rational behavior in terms of maximizing payoffs, and the action of not purchasing information reveals psychological behavior such as fairness.

Since the focus of this experiment is whether Player 1 will purchase information or not, we want to try to isolate this decision as much as possible. To do so, we have to make sure that we can isolate the following three variables that make up this decision: the message that Player 1 sends, the information cost, and the players' preferences over outcomes. By "isolate," I mean control for all factors that might hinder the interpretation of these variables. For example, consider some confounding factors that could occur with particular experiments.

1. An experiment could be conducted in which subjects sit around in a circle and play this game sequentially with each other (a face-to face experiment). This type of experiment would bias analysis of players' decisions to purchase information, since we could not rule out the impact of "seeing other subjects" (see the attraction experiment in Chapter 8). That is, Player 1 might think that Player 2 looks nice and happy and consequently decide to purchase information from him, whereas that same subject might think that another subject looks mean and angry and decide to not purchase information from him. While this could be an interesting experiment, it differs from the previously described experiment, which focuses on the equality of the players' payoff.

2. An experiment could also be conducted that lets subjects communicate with each other in private before making a decision. An experimenter could observe the communication but not the content of the communication (e.g., what they say to each other). The problem here in terms of confounding factors is that the message that Player 1 is supposed to send becomes uncontrolled communication. That is, we cannot interpret the message or know how subjects interpret the message, since we do not know what message has been sent. Hence, we cannot measure its impact. For example, in a private discussion, one player may be more persuasive than another player and may convince her to buy information. But the previous experiment is not about persuasive behavior, so by limiting communication we can control for factors like persuasion.

3. An experiment could be conducted in a survey format in which a survey question describes the problem and then asks the respondent if he or she would purchase information. Let us assume that the survey is voluntary and respondents are not compensated for filling it out. The question is, how can the results be interpreted in terms of our game? They cannot, because in our game, a player can lose something (money) and cause a real inequality (unequal payoffs). In a survey, respondents do not lose anything real (other than time, and since this survey is voluntary, the cost is zero). Because this survey experiment has no information cost, there is thus no inequality to examine. Our game is specifically about these factors. The problem is that in a survey, respondents do not have preferences defined over outcomes or some stake in the outcome, which is needed to establish an information cost.

In order to isolate our three factors or reconstruct our model in the experiment, we could:

1. Control preferences. Ensure that Player 1 has clear preferences regarding outcomes ($0, $1.50, $2.00), and that Player 2 has clear preferences regarding outcomes ($0, $2.00). Subjects should know that these are real amounts and should care about them.

2. Control communication of the message. The messages that players send and receive should be clear and unambiguous so that their impact can be measured across subjects.[10] To place controls on the message, an experimenter could specify that the text

messages sent from one subject to another subject should contain only a single word: king or ace. The simplicity of this message would ensure that all players make the same interpretation of the message. Both players have to understand that the messages are nonbinding to control for the fact that the message is an independent event that impacts beliefs (the likelihood that Player 1 plays a particular card) and does not represent a strategy choice for our players. As we will see later in this book, beliefs and strategy choice are related, but they are two separate elements of a game and must be so in order to define equilibrium behavior.

3. Control the cost of the message. Subjects must understand that information cost is a real cost—that is, Player 1 should know that the outcome of gaining $1.50 is a real, possible outcome. In this case, players will have preferences for higher amounts of money, so taking a reduction of the payoff is a real cost.

4. Control communication among subjects. An anonymous environment should be used as opposed to a face-to-face environment to control for the identity of subjects. Masking subjects' identity controls for any decisions based on a "seeing other subjects" bias. Controls of this sort often mean that barriers have to be erected between subjects so that no leaks (e.g., facial signals, tapping sounds, electronic signals) can occur.

The combination of instituting these controls[11] and randomizing subjects to different treatments allows us to establish causality. In a perfect world, then, the decision to purchase information would allow us to determine whether a subject in an experiment behaved rationally or in terms of some standard of fairness. The problem is that we do not live in a perfect world, and ensuring these strict controls in an experiment is often difficult. Nevertheless, experimentalists attempt to enact controls as tightly as possible. I will come back to this point later in this chapter when I discuss experimental validity.

4. BASELINE COMPARISONS AND CONTROLLING FOR CONFOUNDING FACTORS

Another approach that we use to control for confounding factors is making baseline comparisons, which we mentioned earlier in this chapter. Consider two experiments that use the exact same methods and procedures but differ on a single parameter. Comparisons of the results of the experiments would reveal the impact of that single parameter. If everything else is controlled for (i.e., everything is the same in both experiments), then the parameter would be isolated, allowing easy interpretation of how that single parameter affects results (e.g., whether the parameter causes a negative, positive, or no effect). I should note that experiments often do not allow a unique baseline comparison, and there can be many different baseline comparisons for a single test of a model. For example, in our previous game, some baseline comparisons could be an experiment in which players do not know each others' payoffs, or an experiment in which there is no information cost and a message is sent for free, or an experiment in which no message is sent at all. Many different baseline comparisons could be made, and choosing the correct one depends on the focus of the research question. If the proper baseline comparisons are made, then this method can be an effective way to control for confounding factors.

A central premise of this book is that the laboratory offers the best way of establishing causality in game theory models. Causality entails establishing how one treatment variable impacts another treatment variable or the effect of the independent variable on the dependent variable. In laboratory experiments, as long as subjects are randomized into the different treatments and all parameters are controlled for, it is possible to isolate the effect of one variable on another. That is, by holding all variables or factors constant, casualty between two variables can be measured. Hence, establishing causality through randomization and controls is a powerful approach that experimenters use that is often not available to researchers using observational data. For example, in real-world situations, it is often difficult to control variables such as levels of communication among people who are bargaining or the preferences of individuals who have issues related to fairness.

D. EXPERIMENTAL VALIDITY

1. DIFFERENCES AMONG EXTERNAL, INTERNAL, AND ECOLOGICAL VALIDITIES

One advantage of using experiments is that it is easy to isolate causality because of the simplicity of the experimental environment. The real world is often too complex to analyze because of the interconnectedness of variables involved in an analysis, but experiments allow a researcher to separate out this interconnectedness and pinpoint the significant variables of interest. However, this simplicity has often been criticized. It is often argued that experiments do not reflect the real world, since they are conducted in an experimental laboratory that is devoid of realism. Hence, laboratory experiments are generally criticized for their low validity.[12] In order to understand this argument, we must understand what validity means. Validity is a multifaceted concept, but we will here look at its three main components[13]:

1. *External validity* means that if a particular case (or experiment) is externally valid, then it will extend to other tests (including tests using observational data) to ensure its confirmation. In short, results from a dataset can be replicated across a variety of datasets. For example, external validity can be increased by conducting an experiment in two or more different cultural settings, conducting the same experiment with college students and noncollege students, comparing laboratory results with results from the field, comparing laboratory results with journalistic interpretations of some event, and so on.

2. *Internal validity* concerns how close the assumptions of the model are to the experimental test that is measuring the assumptions, or how well the experiment replicates the model that it is testing. To test internal validity, researchers primarily look at controls in the experiment to determine whether they are instituted the way they are supposed to be instituted. For instance, if it is assumed that subjects are maximizing money, are they being paid enough to ensure that they are maximizing money and not something else? If controls in an experiment are compromised, then the experiment has low internal validity.

3. *Ecological validity* (or *experimental realism*) concerns how closely the methods, materials, and setting of an experiment match some real-world event. For example, how closely does a laboratory election resemble a real-world election?

When most people criticize the generalization of laboratory results to the real world, they are referring to ecological validity. This argument claims that results from a laboratory experiment cannot be used to make inferences about the real world since the laboratory is an artificial environment—that is, a laboratory election is completely different from a real election. But this criticism misses the point. The laboratory election is a real election. Subjects are real voters, and they are electing or defeating real candidates in a situation in which the defeat or election of a candidate has real consequences for voters. The election is different because it takes place in a controlled environment so that a researcher can isolate and measure variables of interest. Unlike real elections, in the laboratory it is possible to induce voter preferences by offering financial incentives and varying preferences and election mechanisms (e.g., plurality vs. approval voting) to determine the impact on election outcomes. In real-world elections it is almost impossible to control the preferences of voters, and voting rules such as approval voting are rarely used. The laboratory affords us the opportunity to examine the impact of these voting rules on outcomes, since the motivation of subjects in a laboratory election should be similar to the motivation of voters in a real election. Plott (1990, 5) notes:

> Economies created in the laboratory might be simple relative to those found in nature, but they are just as real. Real people motivated by real money make real decisions, real mistakes and suffer real frustrations and delights because of their real talents and real limitations. Simplicity should not be confused with reality. Since the laboratory economies are real, the general principles and models that exist in the literature should be expected to apply with the same force to these laboratory economies as to those economics found in the field. The laboratories are simple but the simplicity is an advantage because it allows the reasons for a model's failure to be isolated and sometimes even measured.

2. ARTIFICIAL VS. NATURAL ENVIRONMENTS

By creating an artificial environment, a researcher can concisely measure how subtle changes in the environment impact outcomes. Clearly, researchers would prefer to have total control over some naturally occurring environment, such as an election. For example, assume a researcher is interested in what types of information cause voters to vote for one candidate as opposed to another. Let us assume that a researcher is able to control certain important aspects of a naturally occurring election, such as the information that voters have about the election, and then is able to observe how a voter votes. With these controls in place, the researcher finds a causal relationship between information and voting preferences, but questions still remain about the true nature of the relationship, since he has not controlled for all the confounding aspects of the voting environment. For instance, the researcher would also have to control variables such as the type of voting machine that a voter uses, the time it takes a voter to vote, the lighting and temperature of the polling station, the weather outside, and so on. If these variables are not controlled for, then we cannot rule out the possibility that a voter votes for a particular candidate not because of the information that he or she possesses about that candidate, but because he or she is hot, tired, or using a particular type of voting machine. Again, all researchers would love to have total control over some naturally occurring environment (although such total control would mean the environment was not naturally occurring!), but this is almost always unfeasible. Consequently, the artificial nature of laboratory experiments allows researchers to mimic the naturally occurring environment with total controls to measure behavior within that environment.

3. PROBLEMS WITH EXTERNAL VALIDITY

Given the artificial nature of laboratory experiments, some legitimate arguments can be made against them in terms of external and internal validity. Recall that external validity is assessed on the basis of whether results from a dataset can be replicated across a variety of datasets. Hence, if an experiment finds a particular causal relationship, then the researcher has to determine whether the results hold for a variety of experimental and observational datasets to declare external validity. Since variables in observational datasets usually are not precise enough to test the prediction of game theory models, experimental data are often the most viable method for testing them. But this does not mean that field or observational data cannot also test important aspects of the model. I, like others, encourage testing game theory models across as many different data platforms as possible. Morton (2007, 366) comments:

> It is silly to debate whether data from the lab is superior or inferior to data from the field, either naturally occurring or manipulated. Such analyses should be seen as complements, each telling us something different and unique since each provides us with a different piece of information not available otherwise, which together can provide us with a fuller understanding of the usefulness of the theory.

The problem is that there are often not many different feasible data platforms for a particular research question, and so many models lack external validity. When only experimental data are available, external validity can be increased by conducting the same experiment in a different environment. For example, the same experiment that had been conducted using student subjects from the University of Texas could be replicated using student subjects from Oxford University. This simple comparison would help us to increase external validity by determining if there were any cultural effects observed across the two groups of subjects. That is, this comparison helps us to determine if the results hold when the game is played by two different cultural groups. If the behavior differs across groups, then a model would have to be modified to explain these cultural differences. However, only a few experimental results have been tested across different formulations. Again, as noted in the introduction, laboratory experiments are a new technology, and as experimental results spread, we should also see an increase in the replication of these results in different environments. Here we will discuss a few experiments that have done just that, most notably the ultimatum game.

Also important is the ease with which a researcher is able to duplicate another researcher's dataset. Ease of replication increases a dataset's external validity by proving that the results can be reproduced by a third-party observer. Instructions are usually all that is needed for an experimental replication, but this validity check is more problematic with observational data. Although it is possible to request another researcher's dataset in order to conduct a replication of that study, the data themselves are not regenerated. Unless a researcher is willing to recreate the dataset from scratch (if that is even possible), then the replication will be based on data that may not be clean in the first place.[14]

4. PROBLEMS WITH INTERNAL VALIDITY

An experimental design does not necessarily have to replicate a model exactly; rather, it is supposed to capture some aspect of a model. However, problems arise when the experiment does not capture the particular aspect that it is intended to capture or the assumptions of a model are not replicated properly in the laboratory. For instance, as previously noted, if controls in the laboratory are not instituted correctly, then claims about causality can be biased. If an experimental design calls for no communication among subjects but during the experiment subjects are somehow able to communicate with each other, then the design is flawed and subject to internal validity problems.

There is also a concern in some experiments that the choices that student subjects make in the laboratory do not match the choices that "real decision-makers" have to make. That is, no laboratory experiment conducted on a college campus can represent a decision about an issue such as deploying troops to battle overseas where some deaths will certainly be involved. This is true, but the appropriateness of the subject pool depends on the focus of the model being tested. This would be a concern for experiments that ask subjects to make decisions as if they were an elite or powerful decision-maker. Most game theory models do not focus on this aspect of decision-making, however; rather, they focus on institutional features that shape decisions (see Morton and Williams 2010 for a more in-depth discussion of differences in subject pools).

Finally, some scholars believe that any experiment involving human subjects is fundamentally flawed, since the behavior of subjects within an experimental environment is unlike the behavior of the same subjects in a naturally occurring environment—that is, people behave differently when they know they are being observed. For example, a subject in an experiment might feel pressured to do the "right" or "moral" thing because of the presence of an observer, but he or she might choose a completely different path in a naturally occurring setting. Put differently, actions taken by subjects in an environment where they know they are being observed may differ from actions taken when they know they are not being observed. It is also argued that the experimental environment itself creates an uncomfortable or unnatural environment for subjects. People in the real world do not confront the types of situations that are depicted in experiments, so their behavior in an experiment will always be unnatural (see Levitt and List 2007).

The question, then, is: What is natural behavior? The argument could be made that all behavior that occurs in public is not natural, because everyone in public is being viewed and judged by others and so a person will assume an unnatural persona to deal with these others. If this is the case, natural behavior occurs when a person is alone and in private. Hence, natural behavior is unobservable (unless spy devices are deployed). Consequently, we can only observe people in their unnatural state, and, since the natural state is not observable, we have to accept the unnatural state as the natural one. You could argue that the behavior you see in the real world is closer to a person's natural behavior than a person's behavior in an experiment, but since a person's natural behavior is not known, it could also be the case that behavior in an experiment is closer to a person's natural behavior than his or her behavior in the real world. As well, real-world decision-making often takes place in uncomfortable and unnatural surroundings (e.g., the Mall of America). Since the real world is like this, why shouldn't the experimental environment also be like this?

5. BENEFITS OF AN ARTIFICIAL ENVIRONMENT

Laboratory experiments are currently the best tool that we have to directly test a game's predictions, since the imposition of controls over the experimental environment allows variables to be isolated and studied in a precise way. Although this method is imperfect, it does allow models to be empirically tested so that their implications can be observed and measured. Further testing of a model using different data should always be applauded since it provides more insights about the model.

The benefit of an artificial environment is that it can tell us something about the real world. While laboratory experiments are imperfect, I would disagree with any argument that implies that game theory models conducted in the artificial environment of the laboratory have not taught us anything relevant about human decision-making. Such models have taught us a great deal about human decision-making, as we will see in this book. One of the most exciting parts of experimental research is the researcher's ability to create a little world (such as an entire election system) and manipulate different parameters (different election rules) to determine their impact on the created world.

The goal of experimental research is to use results generated in the laboratory to make important contributions about real-world behavior. The experimental work of Indiana University professor Elinor Ostrom is a prime example. She used game theory and laboratory experiments (among other methods) to study the common pool resource problem (CPR) and ultimately won the Nobel Prize in Economics in 2009.[15] In her acceptance lecture, Ostrom stated that she finds noncooperative behavior to be more likely when people are anonymous, or, in her words, "Isolated, anonymous individuals overharvest from common-pool resources" (2009, 409). Discussing the importance of experiments, she (2009, 426) further noted:

> Experiments on CPRs and public goods have shown that many predictions of the conventional theory of collective action do not hold. More cooperation occurs than predicted, "cheap talk" increases cooperation, and subjects invest in sanctioning free-riders. Experiments also establish that motivational heterogeneity exists in harvesting or contribution decisions as well as decisions on sanctioning.

E. EXPERIMENTAL METHODS

In the social sciences, many different methods are used to conduct experiments, and in this section I will highlight a few different ways in which experiments can be designed.

Psychologists, sociologists, economists, political scientists, and others who conduct experiments often address research questions in different manners, since the questions and theories with which they are concerned are extremely varied. The experimental design often reflects these differences.

In this section I will illustrate the different ways in which experiments can be designed and point out those methods that are used more frequently in certain disciplines as opposed to others.[16]

1. SUBJECT MOTIVATIONS

One distinction among experimental designs is the incentive structure used, with incentives being the rewards a subject receives for agreeing to participate in an experiment. Some experiments pay subjects a flat fee to participate or offer some sort of class credit. A flat fee is generally a fixed amount of money (such as twenty dollars) that is given to participants upon completion of the experiment's task with the understanding that the behavior of the subjects during the experiment will not affect the amount of money awarded.

Other experiments adhere to what is referred to as "induced value theory," a theory partially proposed by Nobel Prize winner Vernon Smith (1976). While it includes different conditions, this theory essentially says that payoffs should be salient to the subjects (i.e., subjects should care about money and believe that more money is better than less money) and the payoffs to the subjects should be tied to their actions (i.e., if they take one action they get one dollar, but if they take another action they get two dollars). In theory, if subjects care about the monetary rewards, then their innate characteristics will be controlled for and they will become rational players and take actions that increase their payoffs or utility whenever possible. Using this method, then, a subject is paid based on his or her performance, so that at the end of the experiment one subject might earn twenty dollars while another subject might earn thirty dollars.[17] It is important to emphasize here that induced value theory is the primary way in which the assumptions of rationality are induced in an experimental environment. This theory stipulates that when subjects value and are focused on the reward medium, their behavior will be consistent with the behavior that is assumed in standard game theory models.

2. DECEPTION

Another difference in experimental designs concerns the use of deception. As the term implies, deception occurs when a subject is given misleading information during an experiment. The Milgrim obedience experiment (1963) is the classic example of deception. In this experiment, a subject was asked to administer an electric shock to another subject who was behind a partition. When the subject pushed the shock button, he could hear another subject scream. The deception in this experiment was that the subject behind the partition was not a subject at all, but a research assistant, who simply screamed whenever the button was pushed. Severe deception of this kind is not possible today because of the creation of human subject rights after World War II, but more benign forms of deception are regularly used in experiments.[18] This methodology is often employed to answer research questions that would be hard to examine without deceiving subjects. For example, if the researcher is interested in studying fear, she might tell the subject that her research concerns trust and then invoke some fear stimuli during the experiment. In this case, if the subjects knew that fear was the purpose of the study, then they could prepare themselves for something fearful to occur and their reaction would not be spontaneous, biasing the results.[19] Other experiments that assume induced value theory do not use deception, because this theory stipulates that subjects should be focused on the monetary rewards in the experiment, and if subjects think they are going to be deceived, then their thoughts will not be on the rewards, but on something else not controlled for by the experimenter.

3. EXPERIMENTAL ENVIRONMENT

The experimental environment can be revealed to subjects in different ways. Some experiments employ scripts, which describe a contextual environment in which they want the subjects to immerse themselves. Scripts take the form of: "Imagine you are the leader of a particular country X and you have control of two-thirds of the population. How would you make a decision Y knowing that two-thirds of the people in the population who support you could be affected?" Other experiments use a generic environment primarily for control issues. In experiments that assume induced value theory, it is felt that a contextual environment might distract from a subject's focus on the main goal of money. For example, in a typical election experiment, two candidates are labeled Candidate A and Candidate B, instead of Democratic Candidate and Republican Candidate, because a subject might personally be a Republican and might vote for a candidate in the experiment because of personal preferences instead of according to the stimuli provided in the experiment.

4. NUMBER OF TRIALS

Experiments can also vary the number of rounds that a game is played. Some experiments use subjects in a one-shot nature—that is, subjects participate in only one round or trial of an experiment. Other experiments repeat a single trial many times. This method is intended to familiarize the subjects with the experimental environment, since some of the experiments can be rather complex. Furthermore, by repeating a single trial, researchers can discover if subjects learn equilibrium behavior over time. Running repeated trials of an experiment allows the experimenter to determine whether subjects' actions will converge to the equilibrium, which cannot be determined by a single trial. When data in this type of experiment are analyzed, it is not uncommon for the researcher to present the last few trials of a repeated trial experiment as evidence of equilibrium behavior. Finally, playing the game repeatedly allows researchers to examine behavior such as reputational behavior, which can only be generated by playing a game more than once.

5. BETWEEN-SUBJECT VS. WITHIN-SUBJECT DESIGN

Differences can also exist in the manner in which subjects are exposed to various treatments. One method is the "between-subject" design, in which a single subject is observed in only one treatment. That is, one subject might be in Treatment A and a different subject in Treatment B. Generally, the subject in Treatment A will not experience Treatment B. Another method is the "within-subject" design, in which a single subject is exposed to multiple treatments. That is, a subject is exposed to Treatment A in the first experimental period and is exposed to Treatment B in the next period. One reason for the difference in designs is the need for some experiments to conduct repeated trials. By randomly assigning subjects to different treatments, the task that subjects are assigned varies each trial (i.e., subjects do not do the same thing over and over). To induce anonymous game play, some experiments randomly match subjects in groups each period, so that they continually play against a different group of subjects and do not know against whom they are playing. Experiments can also use a fixed matching protocol that predetermines how subjects will be matched.

6. ANONYMITY

Another important property of experiments is their choice of an anonymous or face-to-face design. In a face-to-face experiment, subjects usually know their opponents, since they are physically close to each other, and sometimes know how another subject has responded to a query. In an anonymous environment, subjects often play the game on computers, and although they may be physically close to each other, they cannot determine how another subject has responded to a query. In this environment, a subject does not know his or her opponent. In many experiments, it is important to conceal the identity of a subject so that he or she can make responses without fear of reprisal outside of the experimental laboratory. In single-blind experiments, the identity of a subject is hidden from other subjects but not the researcher.[20] In double-blind experiments, neither the researcher nor the subjects know each other's identity. Experiments use anonymity because if a subject is concerned with reprisal from other subjects after the completion of the experiment, then her decisions during the experiment might be compromised or biased. However, anonymity presents a case in which subjects do not know what decisions other subjects have made, and so there should be no concerns about reprisals.[21]

7. HOW DO YOU DESIGN A GOOD EXPERIMENT?

The previous sections have outlined only a few of the fundamental design issues that can be implemented in an experiment. Again, the research question in hand determines what design will best establish causality. New technology will no doubt further enhance innovative designs, adding to my list of fundamentals. You might ask: How do I design a good experiment? There is no magic formula for how to design an experiment, since each design is specific to the particular problem being studied. However, for game theory experiments, the experimental design is specified by the model that is being tested. Experimenters attempt to recreate the model's assumptions in the laboratory as precisely as possible. Because it is often difficult to recreate the precise model in an experimental environment, experimenters attempt to design simple experiments that test some relevant assumption or prediction of their model. Friedman and Sunder (1994, 10–11) suggest:

Your goal should be to find a design that offers the best opportunity to learn something useful and to answer the questions that motivate your research. Usually an effective design is quite simple compared to reality, and in some respects simpler than relevant formal models. . . . It is futile to try to replicate in the laboratory the complexities of a field environment. . . . No matter where you stop in building the details of reality into your laboratory environment, an infinite amount of detail will always remain uncaptured. . . . Before you get close, the laboratory environment will have become so complex that you will find it difficult or impossible to disentangle causes and effects. As in any other experimental discipline, simplicity enhances control. Try to find the simplest laboratory environment that incorporates some interesting aspects of the field environment.

While no handbook can describe how to design an experiment, I think that Friedman and Sunder's suggestion of simplicity is the best advice. A simple experiment that yields relevant results, like the Franklin experiment, is usually the best experiment.

F. SUMMARY

In this chapter I have provided an overview of experimental methods, emphasizing that the strength of experiments lies in their ability to allow a researcher to establish causality between the variables of interest. By controlling all aspects of the experimental environment and randomizing subjects to different treatments, it is possible to isolate the effect of one variable upon another. This is often difficult to do using observational data, since many factors may not be controlled for when the data are generated. I have also outlined the different methodologies that can be used in an experiment. The type of method used depends on the research question that is being addressed. In the next two chapters we will examine utility theory, which posits the behavioral assumptions of players in a model and forms the foundation of game theory. Chapter 3 will discuss ordinal utility and Chapter 4 will discuss cardinal utility.

NOTES

1. Although this awkward term does not describe the all-encompassing nature of the experimental endeavor, I use it because it is the common term used to describe these types of experiments.
2. See Appendix 2 for a history of political economy experiments.
3. Say's Law refers to Jean-Baptiste Say, a French economist who lived from 1767 to 1832.
4. One such program is z-tree (see http://www.ztree.com/).
5. The mean is usually squared to provide a more accurate estimation.
6. However, it has been pointed out to me that Franklin's experiment might have randomized the clothes to different treatments, as long as the dark fabric was assigned to one patch of snow and the light fabric was assigned to another patch of snow. As long as the fabric was randomly assigned to equivalent patches of snow, then inferences could be made based on the different treatments. However, Franklin left no mention of his methodology, so we cannot discern what procedures he used.
7. Researchers who do field experiments rely entirely on randomization to establish causality, because controls are difficult to establish in the field.
8. This example is similar to an experiment by Roth (1995a) that is discussed in Chapter 9.

9. There is also an important difference between observable and unobservable confounding factors when one purpose of an experiment is to make unobservable factors observable by instituting controls (see Morton and Williams 2010).

10. That is, every subject in the experiment (assuming multiple trials are conducted) should interpret the message in the exact same way so we can compare behavior relative to the message.

11. As well as other controls, such as making sure there are no distractions in the laboratory.

12. I have often heard the statement "experiments have high internal validity but low external validity." I have never quite understood what this means, because it incorrectly defines the concept of external validity.

13. This is just a bare-bones distinction. The study of validity is actually very complex, and I am greatly simplifying the concept here. For instance, internal validity involves causal, statistical, and construct validity. External validity, of which ecological validity is a part, is also defined by statistical replication. See Morton and Williams (2010) for a more detailed discussion of validity.

14. As an experimentalist, every time I evaluate some observational data I always ask the person from whom I get the data, "Where did these numbers come from?" Which consistently generates an odd look.

15. Vernon Smith, one of the founders of the experimental method in economics and a 2002 winner of the Nobel Prize, is another good example.

16. For a fuller discussion of the differences in experimental methods, see Hertwig and Ortmann (2001).

17. In addition, experimenters usually pay subjects a "show-up fee" to appear on time for the experimental session. This fee is usually independent of any money earned during the experiment or for time spent participating in an experiment.

18. See Morton and Williams (2010) for a history of human subject rights and the development of institutional boards that protect the safety of human subjects.

19. Some experiments also debrief subjects about the purpose of the study immediately after the experiment. Explaining the purpose underlying the experiment is supposed to serve as an educational asset and eliminate any "harm" that subjects might feel about an experiment's deceptive nature or other features that might impact the subject's psyche (see Morton and Williams 2010).

20. Blindness can also mean that subjects are unaware of the treatment that is being administered.

21. In some experiments, it is customary to pay subjects in private so that they do not know how much other subjects have earned. Usually subjects are told prior to the experiment that they will be paid in private to ensure that decisions in the experiment are not biased by payments. Good experimental labs have two rooms to allow the payment exchange to be conducted in private.

CHAPTER 3

ORDINAL UTILITY THEORY

A. TOO MANY CHOICES?

This chapter is concerned with preferences and choice. As Schwartz (2004) comments, preferences and choices in everyday life are not simple calculations. Looking at consumer choice, he observes (12):

> A typical supermarket carries more than 30,000 items. That's a lot to choose from. And more than 20,000 *new* products hit the shelves every year, almost all of them doomed to failure.
>
> Comparison shopping to get the best price adds still another dimension to the array of choices, so that if you were a truly careful shopper, you could spend the better part of a day just to select a box of crackers, as you worry about price, flavor, freshness, fat, sodium, and calories. But who has the time to do this? Perhaps that's the reason consumers tend to return to the products they usually buy, not even noticing 75% of the items competing for their attention and their dollars. Who but a professor doing research would even stop to consider that there are almost 300 different cookie options to choose among?
>
> Supermarkets are unusual as repositories for what are called "nondurable goods," goods that are quickly used and replenished. So buying the wrong brand of cookies doesn't have significant emotional or financial consequences. But in most other settings, people are out to buy things that cost more money, and that are meant to last. And here, as the number of options increases, the psychological stakes rise accordingly.

What this comment illustrates is that choice over alternatives that have little financial consequences can be problematic when there are numerous alternatives from which to choose. The rational actor we study can easily have preferences defined over numerous alternatives, but usually only two or three alternatives are needed to analyze a wide range of important decision environments. However, for behavioral game theorists, choice over many alternatives could be relevant to study how people narrow the range of alternatives to a few manageable choices when they are confronted with many alternatives. One way people do this is by using a decision heuristic or information shortcut, such that when a person is confronted with 300 different cookie choices, the number of feasible alternatives is reduced by only considering cookies that are on sale, prominently displayed, or a particular brand. To illustrate the utility of this type of heuristic to filter through alternatives, imagine that a friend gives you an iPod with one thousand albums. How would you decide which song to play first or which songs to play in the first half hour of listening?

B. STRICT RATIONALITY

Strict rationality concerns assumptions about a person's preferences and choice of alternatives. It assumes that a person can rank various alternatives and will select the best available alternative; in short, a person will maximize utility. Rationality does not assume self-interested behavior, but it does acknowledge that when people in a game situation make choices that yield themselves the highest available utility, these choices may have a detrimental impact on other people in the game. In a long definition of rational choice theory, Osborne (2004, 4) states:

> The theory of rational choice is a component of many models in game theory. Briefly, this theory is that a decision-maker chooses the best action according to her preferences, among all the actions available to her. No qualitative restriction is placed on the decision-maker's preferences; her "rationality" lies in the consistency of her decisions when faced with different sets of available actions, not in the nature of her likes and dislikes.

In a shorter definition, Osborne summarizes rational choice as "the action chosen by a decision-maker is at least as good, according to preferences, as every other available action" (2004, 10).

One reason that models assume rationality is that in order to solve games, we need a consistent theory of human behavior. If we were to assume that players behaved randomly, then it would be impossible to predict how a player would play in a particular game situation. The assumptions of rationality allow preferences to be modeled mathematically. To represent preferences mathematically, certain consistency conditions must be met that are fulfilled by assuming rationality.

Morrow (1994, 20–21) provides four misconceptions about the notion of rationality that I find useful to quote here:

1. We do not assume the decision process is a series of literal calculations. Instead, people make choices that reflect both their underlying goals and the constraint of the situation, and we can create a utility function that represents their actions given those constraints.
2. Rationality tells us nothing about an actor's preferences over outcomes—only about its choices given those preferences and the situation that confronts it.
3. Rational actors may not and probably will not all reach the same decision when faced with the same situation. Rational actors can differ in their preferences over the outcomes.
4. Rational actors can make errors, that is, achieve undesirable outcomes, for three reasons....First, situations are risky....Second, the information available to actors is limited....Third, actors may hold incorrect beliefs about the consequences of their actions.

We should keep these misconceptions in mind when we study utility theory in this and the next chapter.

Utility can be represented in two forms: ordinal utility and cardinal utility. Ordinal utility concerns the rankings of alternatives. If A is preferred to B, no judgment is made regarding how much A is preferred to B and no statement is made about intensity of preferences. Cardinal utility, in contrast, can measure intensity of preferences. In this chapter, we will discuss ordinal utility and move on to cardinal utility in the next chapter. In the next section, I will discuss utility theory and the properties of ordering alternatives. In the fourth section, I will define an ordinal utility function and then illustrate a spatial utility function. In the fifth section, I will show how utility functions are induced in laboratory experiments. Finally, in the last section, I will explain social preferences and modeling emotions in preferences, which will provide us with a framework to understand the experiments discussed in the rest of the book.

C. UTILITY THEORY

1. UTILITY

Utility is a unit of good, or "utile," for anything that a person values, such as money, food, fast cars, a large home, a high-definition TV, a certain candidate, and so on. Davis defines a utility function as "a 'quantification' of a person's preferences with respect to certain objects" (1982, 62). Utility functions connect actions (or strategy choices) to outcomes (payoffs), or map values into utility. I would like to emphasize that people do not have utility functions per se, but they do have preferences that can be represented by utility functions. It is assumed that a set of actions A, or a strategy set, is a finite number, $A = (a_1, a_2, \ldots a_n)$. This means there are not an infinite number of strategies (which could make models insolvable). Hence, for a voter in a two-candidate election, A = (vote for Candidate Z, vote for Candidate Y, abstain). Furthermore, it is assumed that the set of outcomes is also finite, with outcomes $X = (x_1, x_2, \ldots x_n)$. In the election case, then, X = (Candidate Z wins, Candidate Y wins, tie). A utility function, u, maps (or converts) actions (A) into an outcome (X): $A \rightarrow u \rightarrow X$. In this example, a utility function would define for a voter preferences for Candidate Z, Candidate Y, or neither.

2. GRAPHICAL UTILITY FUNCTIONS

To informally introduce the idea of how a utility function transforms some value into utility, I will first introduce a graphical presentation of this concept. Consider Figure 3.1.

Figure 3.1 measures utility on the vertical axis and the value of some good (say money) on the horizontal axis. The linear line within the graph is a utility function that measures how much utility is derived from a specific value. Let us assume that utility is measured in terms of money, so that 500 dollars is translated to a utility of 50, u_i ($500) = 50 utiles, indicated by the dashed line. This simply means that the utility for person i (the subscript) of 500 dollars is 50 utiles. Also notice that according to the dashed line, u_i ($800) = 80 utiles. This linear utility function indicates that a person always prefers more money to less. Hence, for every increase in monetary value there is also an equal increase in utility. However, in certain situations, this utility function is a bad model of human behavior. If we were to construct a graph with axes extending to 500 billion dollars, it is likely

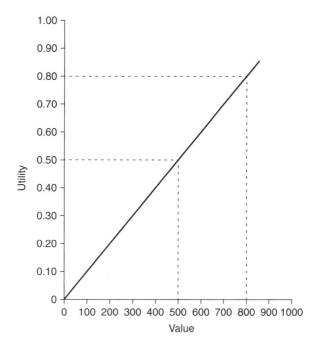

FIGURE 3.1 Linear utility function

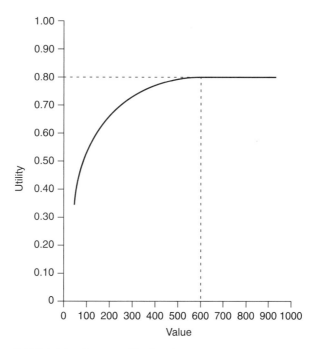

FIGURE 3.2 Concave utility function

that at a certain point, a person would not value 200 billion dollars and one cent more than she would value 200 billion dollars. At that point, these monetary amounts would become the same—a player would cease to gain satisfaction or utility from small increases when billions of dollars were involved. In other words, there would be a marginal decreasing return for units of utility. If we are thinking about utility in terms of money, then a better utility function would be like the concave function displayed in Figure 3.2.

In this case, a person values money up to a certain point, at which utility for the values stays the same. The dashed lines in the figure represent the utility for 600 dollars, or u_i ($600) = 80 utiles, as the point at which the person stops getting increased utility from money. In this region of the utility function, a player becomes indifferent to values, since she receives the same utility for each increasing value.

Finally, think about the utility for beer on a graph where the horizontal axis represents the number of beers consumed and the vertical axis represents the utility for beer. According to this graph, the first two beers are enjoyable, but after the sixth beer the experience begins to have negative consequences and the line begins to slope downward.

Notice that the utility for six beers is 80 but the utility for eight beers is around 62. Utility functions allow us to convert values into utility and model different types of behavior by changing the shape of the function or just bending the line.

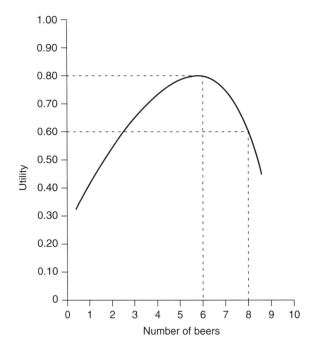

FIGURE 3.3 Utility for beer

D. ORDERING ALTERNATIVES

1. RESTRICTIONS ON CHOICE

According to utility theory, people can have utility for anything, but they must be able to rank their preferences for alternatives. First, let us consider two important restrictions regarding the ordering of alternatives.

Connectivity (comparability): Given two outcomes x and y, either x ≥ y or y ≥ x.

This means that for two ordered pairs, a player is indifferent between x and y or prefers y to x.[1] Connectivity means that players can compare all available outcomes and determine that they prefer one to another or are indifferent among all outcomes. This concept rules out the case in which a player cannot determine the best outcome because of an inability to make comparisons.

Transitivity: If x ≥ y and y ≥ z, then x ≥ y ≥ z.

This means that for three ordered pairs, if you prefer x to y and y to z, then you can rank your preferences for the three alternatives: you most prefer x, then you prefer y, and you least prefer z. You can also be indifferent among three alternatives. These preferences are called transitive preferences; in other words, if you prefer red to blue and blue to pink, then, by logical extension, you prefer red to pink. If individual preferences are transitive, then outcomes can be ordered whenever there are more than two alternatives; in the case of color preferences, we can say that an individual most prefers red ≥ blue ≥ pink. In this case, if a player had to make a choice among three alternatives and red was available, then he would select red. Intransitivity would be the case in which red is preferred to blue and blue is preferred to pink, but pink is preferred to red. In this case, red ≥ blue ≥ pink ≥ red. This statement implies that pink is preferred to blue, but we have already stipulated that blue is preferred to pink, producing an intransitive preference ordering and making the preference that stipulates that pink is preferred to red inadmissible.

Both of the assumptions of connectivity and transitivity allow players to order outcomes. Hence, when presented with a set of outcomes, we assume that players will behave rationally and rank outcomes from best to worse, knowing that they might be indifferent to some outcomes.

2. MAY'S INTRANSITIVE PREFERENCES EXPERIMENT

May (1954) conducted an experiment in which male college freshman math students were presented with three hypothetical marriage partners, x, y, and z. Each potential mate differed on three dimensions: in intelligence, x > y > z; in looks, y > z > x; and in wealth, z > x > y. Subjects were told that Partner x was very intelligent, plain, and well-off; Partner y was intelligent, very good-looking, and poor; and Partner z was fairly intelligent, good-looking, and rich. Subjects were then asked to order the three alternatives, with indifference among the alternatives not being allowed. The results showed that x > y, y > z, and z > x. The resultant ordering was thus intransitive: z should not beat x, since x beats y and y beats z. Commenting on these results, May notes that the experiment "does not prove that individual patterns are always intransitive. It does, however, suggest that where choice depends on conflicting criteria, preference patterns may be intransitive unless one criterion dominates" (1954, 7).

May also points to another psychology experiment in which hungry rats were shown to have the following preferences: food was preferred to sex, sex was preferred to the avoidance of pain, and the avoidance of pain was preferred to food[2]—another intransitive ordering.[3] What is important to note from these experiments is that the assumption of transitivity does place restrictions on the preference order that a player is allowed to have. I will illustrate this point in more depth later.

3. CHOICE AND TIME

Consistency of preferences is time dependent in the sense that the preferences apply at the moment a player makes a decision. For example, a player may have a certain preference for alternatives at time t (the present), but the same player may have a different preference ordering for the same alternatives at time t + 1 (the future). Schuck-Paim, Pompilio, and Kacelnik (2004, 2306) comment:

> It is worth remembering that consistency of preference is accepted by all parties to be only relevant when constancy in the state of the subjects and in the properties of the options is assumed. A subject that prefers lamb to ice cream before dinner, ice cream to coffee immediately after dinner, and coffee to lamb a few minutes later is not considered to be showing intransitivity or violating any principle of rationality, because she is (trivially) changing states between choices.

4. NONPERVERSE SELECTION RULE AND EXHAUSTIVE SET OF ALTERNATIVES

In addition to the two ordering assumptions described previously, other assumptions must be made in order to define rationality. First, it is assumed that players select their highest-ranking alternative given their current environment. If a player has a preference for A > B > C and has an opportunity to select A, she will select A; if the environment prevents her from selecting A, she will select B if possible. This assumption is based on a nonperverse selection rule, which specifies that the player will not select C given a chance to select A. For example, say I offer you a Coke or a Pepsi and you reply, "I prefer Coke, so give me a Pepsi." Your choice is perverse, because we assume that the preferred alternative is the one chosen. The nonperverse selection assumption means that if an individual has preferences that are tied to choices, then the individual will choose the best available outcome.

Second, it is assumed that an exhaustive set of choices exists. That is, the only choices available are the ones specified in the game. If a player is given a choice between A and B, then that player cannot choose C. Thus, the choice set is exhaustive and players must make choices from among that set.

5. ARIELY'S *ECONOMIST* EXPERIMENT

This exhaustive set of choices assumption is nontrivial, since additional choices can often change the choice environment. Consider the example given by Ariely (2008), in which the *Economist* made the following offer for a subscription to their newspaper and website. The offer was as follows:

1. One-year subscription to *Economist.com* for $59 (website only)
2. One-year print subscription to the *Economist* for $125 (print only)
3. One-year print subscription and one-year web subscription for $125 (both website and print)

The fact that Options 2 and 3 are the same monetary amounts is no mistake. Ariely argues that this strategy is a ploy to get people to subscribe to Option 3 by adding a "decoy" Option 2. He (1998, 2) comments:

> I may not have known whether the Internet-only subscription at $59 was a better deal than the print-only option at $125. But I certainly knew that the print-and-Internet option for $125 was better than the print-only option at $125. In fact, you could reasonably deduce that in the combination package, the Internet subscription is free! "It's a bloody steal—go for it governor!"

Ariely ran experiments comparing the original options against a set of modified options that deleted Option 2:

1. One-year subscription to *Economist.com* for $59 (website only)
2. One-year print subscription and web subscription for $125 (both website and print)

He found that when the subscription offer had three alternatives, including the decoy alternative, 16 subjects chose Option 1, 0 subjects chose Option 2, and 84 subjects chose Option 3. When the decoy offer was omitted, 68 chose Option 1 and 32 chose Option 2. Hence, the addition of a decoy option changed the preferences of subjects. When considering our games, it is important to understand how the available choices affect game play and how alternative choices might alter equilibrium play.

E. ORDINAL UTILITY FUNCTIONS

An ordinal utility function for a player i, defined as u_i, simply assigns a number to some outcome, with the numbers ranking each outcome from best to worse. For example, say that a player has three outcomes, O_1, O_2, and O_3, where O_1 is an orange, O_2 is a grape, and O_3 is a peach. We can assign a number to each outcome. Say that we assign a 5 to the orange, a 1 to the grape, and a 3 to the peach. This utility function can be represented as follows: $u_i(O_1) = 5$, $u_i(O_2) = 1$, $u_i(O_3) = 3$. This "fruit" utility function represents the preferences $O_1 > O_3 > O_2$, indicating that an orange is preferred to a peach and a peach is preferred to a grape. The assignment of (5, 3, 1) is arbitrary, since only the order of the numbers is important. Hence, we could say that $u_i(O_1) = 100$, $u_i(O_2) = 0$, and $u_i(O_3) = 1$, and this would convey the same ordering.

A player is indifferent between two outcomes if the utility is the same. So if $u_i(O_1) = 5$, $u_i(O_2) = 3$, and $u_i(O_3) = 3$, then the player is indifferent between the second and third outcomes. In general, in a choice between two outcomes X_j and X_k, $u_i(X_j) > u_i(X_k)$ if and only if X_j is preferred to X_k.

As the introduction to this chapter notes, ordinal utility numbers do not model *intensity* of preferences, but only provide a ranked ordering among the various alternatives. Consequently, there is no way to compare two individuals' intensity of preferences, which is referred to as interpersonal comparisons of utility. So, if one person's utility for O_1 is $u_1(O_1) = 5$ and another person's utility for O_1 is $u_2(O_1) = 10$, this does not imply that the second person has double the utility for O_1 than the first person. These utility values only represent individual rankings.

F. SPATIAL PREFERENCES IN ONE DIMENSION

1. MODELING IDEOLOGY

Other types of ordinal utility functions can be represented graphically, as in Figures 3.1 and 3.3. Figure 3.4 presents a different shape, which is called a spatial utility function and is defined over a single dimension. Typically, these utility functions are restricted to one or two dimensions.[4] Figure 3.4 illustrates a typical one-dimension utility function.

The value dimension in this figure can be labeled anything, but for the purpose of illustration let us assume that the dimension represents an ideological spectrum, where the extreme left represents extreme liberal policy, the middle represents moderate policy, and the extreme right represents extreme conservative policy. All points in between are increments of these positions. Notice that the single highest peak of this utility function occurs at its moderate (middle) position. This peak is called a player's ideal point and

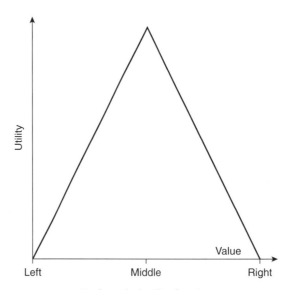

FIGURE 3.4 Single-peaked utility function

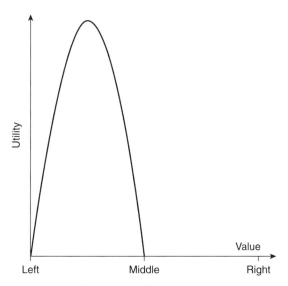

FIGURE 3.5   Skewed utility function

represents the position on the dimension that she most prefers and at which she derives her highest utility. As we descend monotonically away from the ideal point, the player receives less and less utility. Note that the player will be indifferent between any two points that are equidistant from the ideal point, since they will both yield the same utility. Hence, she derives the same utility from an extreme left position as she does from an extreme right position. If you think that this conclusion does not make sense, you are correct. If we were concerned with modeling a person's ideology, we would probably use a different utility function, like the one depicted in Figure 3.5. Spatial utility functions can take on many shapes. Figure 3.5 depicts a liberal who gets very little utility from conservative positions.

2. SINGLE-PEAKEDNESS AND TRANSITIVITY

Spatial utility functions must satisfy one condition: they must contain a single peak. This restriction, called single-peakedness or satiable preferences, is the same as the transitivity assumption discussed previously. This assumption means that there cannot be a dip in the utility function, as in Figure 3.6. A dip indicates that a person has two ideal points, and so we cannot define his or her preferences, since they are not transitive.

To see how this assumption about single-peakedness can restrict the set of allowable preferences, consider the game reported in Strom (1990). In this game, three committee members must vote on whether to report a bill (Report Bill), amend a bill (Amend Bill), or send the bill out of committee (No Bill). The members vote for alternatives in a pairwise comparison; that is, they first vote on whether to report the bill or amend the bill, and then pit the winning alternative against the no-bill alternative. Assume that the committee members have the following preferences:

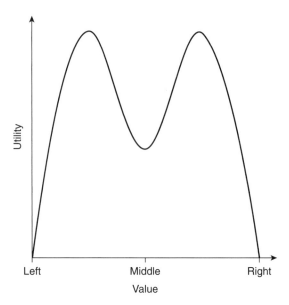

FIGURE 3.6 Utility function with a dip

Member A: Report Bill > Amend Bill > No Bill
Member B: Amend Bill > Report Bill > No Bill
Member C: No Bill > Report Bill > Amend Bill

Now we can graph these preferences spatially in Figure 3.7. Notice that Member A's first preference is Report Bill, which corresponds to the highest position in the figure (first preference); her second preference is Amend Bill, which corresponds to the second highest position in the figure (second preference); and her least preferred alternative is No Bill, which corresponds to the lowest position in the figure (third preference).[5]

In this example, all members have single-peaked preferences, depicted by the "unique" high point that appears on each line in the graph. In a vote between Report Bill and Amend Bill, Report Bill wins, since Members A and C both prefer Report Bill to Amend Bill. In a vote between Report Bill and No Bill, Report Bill wins again, since Members A and B both prefer Report Bill to No Bill. Given these preferences, Report Bill would ultimately be the winning alternative.

Now let us alter Member B's second and third preferences as follows:

Member A: Report Bill > Amend Bill > No Bill
Member B: Amend Bill > No Bill > Report Bill
Member C: No Bill > Report Bill > Amend Bill

The graph of Member B's new preferences, as shown in Figure 3.8, no longer has a single peak. There is no unique high point.

Now, in a vote between Report Bill and Amend Bill, Report Bill wins; in a vote between Report Bill and No Bill, No Bill wins; and in a vote between No Bill and Amend Bill, Amend Bill wins. Hence, there is no clear winner and the social ordering is intransitive. In this example, the assumption of transitivity would rule Member B's preferences inadmissible, since they are not single-peaked.

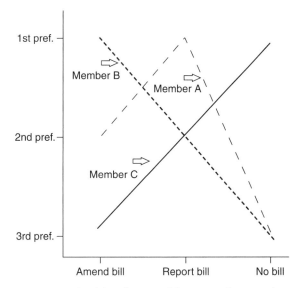

FIGURE 3.7   Spatial preferences of three committee members

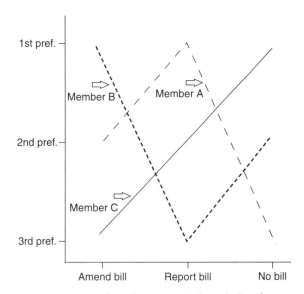

FIGURE 3.8   Member B has a non-single-peaked preference

G. HOW UTILITY FUNCTIONS FOR MONEY ARE INDUCED IN POLITICAL ECONOMY EXPERIMENTS

1. PAYOFF CHARTS

As noted in Chapter 2, experiments that adhere to induced value theory pay subjects money based on their performance during the experiment. Sometimes utility functions in these experiments are assumed, as in the case when two players bargain over some dollar amount or bid in some auction. Other times a utility function that determines how much money

TABLE 3.1 PAYOFF MATRIX FOR GREEN VOTER

	G wins election	R wins election
Vote G	$1.50	$0.00
Vote R	$0.25	$1.00

Source: Reprinted with permission from Springer Science and Business Media; Dasgupta et al. 2008.

a subject can earn during the experiment is formally defined.[6] Depending on the nature of the study, a wide range of utility functions can be used. Here I will discuss two methods by which utility functions are assigned: a payoff matrix and spatial preferences. In the payoff matrix method, subjects are given a table that associates outcomes to a monetary amount. Table 3.1 is an example of such a matrix.

In many experiments, subjects are assigned a type (this will be discussed further in Chapter 10). As noted in the last chapter, experiments that assume induced value theory use a sterile experimental environment. Hence, voting games in such experiments use a nonpolitical environment, so that subjects vote for Green, Red, and Blue candidates (for example) instead of Democratic or Republican candidates. Subjects themselves are assigned a color type that identifies the candidate whom they most prefer. So Green voters most prefer Green candidates. This does not necessarily mean, however, that Green voters will have an incentive to vote for the Green candidate.

This example represents the utility function for a subject assigned to the Green type. According to the matrix, if he votes for the Green candidate and this candidate wins, then he will receive $1.50. However, if he votes for the Green candidate but the Red candidate wins, he will receive no money for this election period (his worse outcome). If he votes for the Red candidate and the Green candidate wins, he receives a quarter for that election (his second worst outcome). Notice that his second best outcome is when he votes for the Red candidate and the Red candidate wins. Using this utility function, then, if the subject thinks the Green candidate will win, he has an incentive to vote for that candidate, but if he thinks the Red candidate will win, he has an incentive to vote for the Red candidate. As we will see in this book, variants of this sort of payoff matrix are used in many types of experiments.

2. SPATIAL PAYOFFS

Another method for assigning utility functions is to use a spatial utility function, like the one in Figure 3.9.

This figure can be used to illustrate an election in which two candidates are competing over the x dimension by selecting issue positions from 0 to 1,000. The candidate who wins then selects a position on the dimension, say 200. A monetary payoff is given to the subject in accordance with the position selected. In this example the subject's ideal position is 700, at which point he gets 95 cents. As a result, he prefers candidates to adopt winning positions

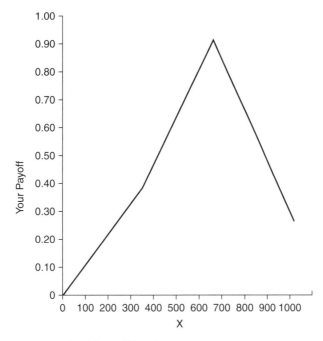

FIGURE 3.9 Spatial payoff function

Source: Reprinted with permission from Elsevier; McKelvey and Ordeshook 1985.

that are close to his ideal points. Since the winning candidate adopted 200, then this subject would only receive 20 cents.[7] Notice that in neither the payoff matrix nor the spatial utility function example were payoffs substantial. However, as noted in the last chapter, a single trial of an experiment might be repeated a number of times, allowing subjects to collect the accumulated money earned in all trials, which has the potential to be substantial.

H. RATIONALITY, EMOTIONS, AND SOCIAL PREFERENCES

1. RATIONALITY AND EMOTIONS

Rationality by itself does not involve emotions. Instead, it dictates that in order to make an optimal selection, all that a player has to know are her preferences over some alternatives. This decision does not entail emotions such as anger or happiness that might alter "rational" decisions. Wilson (2005, 9) comments:

> It is no surprise that emotions affect behavior. After all, human emotion is a part of our evolved nature. Indeed, emotion is a critical part of how humans (and other animals) navigate daily life.... Emotions systematically affect the ways in which people evaluate information, the ways in which they assign risk to an event and the manner in which they choose among simple strategies.

A game theory model can incorporate emotions by basing them on a player's knowledge about his own payoffs and/or the payoffs of other players. In game theory, four payoff-relevant cases can occur.[8]

1. A player can unexpectedly receive an increase or decrease in her own payoff.
    Possible emotions involved: joy, anger, sadness, surprise, kindness, anticipation, relief, trust
2. A player can forecast the deletion or addition of payoff-relevant alternatives.
    Possible emotions involved: joy, sadness, anger, surprise, relief, trust
3. A player can have her payoff increased (or decreased) as a result of actions from another player or some external enforceable social planner.
    Possible emotions involved: fear, sadness, joy, kindness, disgust, anger, revenge, relief, trust
4. A player can have a payoff that is higher, lower, or the same than that of another player.
    Possible emotions involved: envy, disgust, sadness, joy, anger, jealousy, revenge

Consider our example in Chapter 1 about a person who goes to a restaurant to order food. Assume that one day he goes and orders a hamburger only to find out unexpectedly that the restaurant has run out of them (Case 1). Rationality dictates that our person will "neutrally" settle for his second or third preference. But what happens if our person is so upset that the restaurant has run out of his favorite food that he angrily refuses to acknowledge his other choices and leaves the restaurant? Is this behavior irrational? In the strict sense of the term, this behavior violates assumptions of rationality, because in this framework, if a first choice is not an option, then players are required to make a choice among lesser alternatives. We can imagine situations in which this behavior would be deemed rational if an emotion such as anger were considered. For instance, it can be assumed that when a person's feeling for an alternative is so strong (here, his lust for a hamburger), the deprivation of this alternative will produce such a feeling of anger in him that it will diminish (or eradicate) his desire for other, lesser alternatives.

2. RATIONALITY USED TO STUDY OTHER TYPES OF BEHAVIOR VIA DEVIATIONS

Although rationality is criticized for being an oversimplified version of human behavior, it is this oversimplification that allows us to test other aspects of human behavior such as emotions. Rationality provides a benchmark for researchers to test how a particular decision-making model compares with a rational decision-making model in terms of predictive behavior. For instance, assume a game with two players. The first player is given 10 dollars and told that she can give the second player five dollars and not receive anything in return, or else she can keep the 10 dollars and leave the experiment. Rational choice theory assumes that the first player should not willingly accept a reduction in utility (especially in the short term), so she should choose to leave the experiment with 10 dollars. But in cases in which our first player willingly gives the second player five dollars and expects nothing in return, it is clear that she is not acting in a self-regarding manner. This behavior represents a deviation from the predicted behavior—that is, in terms of rationality, this decision is a mistake. Consequently, if we can rule out self-regarding behavior as the reason for this player's decision, then we must find some other behavioral explanation. It is possible to build models about particular motivations, such as an emotion like kindness, that could explain this behavior. In this example, the emotion for kindness could have "crowded out" self-regarding emotions, leading the player to forgo half of her payoff and expect nothing in return.[9] While this is a reasonable explanation, it is unfortunately not the *only* explanation, since other emotions, processes of emotions, or even nonemotional factors could predict and produce the same behavior. So the question becomes: Which model of behavior provides the best empirically sound explanation?

When an experiment is conducted, the experimenter, for the most part, does not observe the decision-making process but only the decision outcome—that is, the experimenter does not observe the processes involved in the deviation, only results of the deviation. Consequently, many models can be created to explain behavior, because deviations can be the result of many factors. Consider subject error. Anyone who has conducted a reasonable amount of experiments will understand that a small number of subjects will not understand their task in the experiment, no matter how careful the experimenter is in explaining the instructions. As well, some subjects will become bored with their task and select responses randomly to pass the time. Other subjects will try to figure out the purpose of the experiment and outsmart the experimenter, resulting in deviations. These experiment-related mistakes on the part of subjects cause some of the deviations in results. However, even when an experimenter controls for simple subject error, systematic deviations still occur, indicating that the deviations form a pattern of behavior that is nonrandom or the "mistakes" (in terms of the assumptions of rational choice theory) are not really mistakes, but behavior that can be explained. Behavioral game theorists argue that these deviations are caused by emotions not controlled for in the model or experiment. They assume that a person's utility function has both a self-regarding component and an emotional component. This emotional component is referred to as social preferences.

3. SOCIAL PREFERENCES DEFINED

It is difficult to provide a precise definition of social preferences, because they tend to fall into the category of "I know it when I see it!" or "I know something is there because I can touch it and feel it, but what is it?" Loosely speaking, this term means that a person cares about another player's payoff (prior cases 3 and 4). In other words, a person cares about his own payoff in terms of maximizing utility but is also concerned with another player's payoff in terms of maximizing behavior. Recall that this motivation violates the assumptions of rationality and induced value theory, since a subject who is sufficiently motivated should be concerned about only his payoff and not the payoff of other players. Behavioral game theorists have refined the strict utility-maximizing model to include this social element. The following lists concepts these theorists use when referring to social preferences[10]:

- *Self-regarding preferences*: A person is only concerned with his or her payoff. This is the standard assumption of rationality.
- *Other-regarding preferences*: A person is concerned with not only his or her payoff, but also the payoff of another person. Sometimes this behavior is referred to as "pro-social

behavior," or an individual's voluntary actions to help or benefit one or more individuals. Pro-social behavior is concerned with a pattern of activities. Types of pro-social behavior include altruism, or unselfish interest in helping one or more individuals, and empathy, or caring for others. When an individual engages in a series of acts of altruism or empathy, he or she is engaging in pro-social behavior. Other-regarding preferences refer to a case in which a person is concerned (for whatever reason) about the welfare of others.

- *Inequity-averse social preferences (fairness preferences)*: A person is concerned with the equality of her payoff in relation to another person's payoff. That is, an individual has an aversion to an unequal distribution of payoffs, especially when the distribution is perceived to be unfairly biased. In this case, a person's knowledge about the payoff of another person matters when there are perceived disparities in the distribution of payoffs. This social preference invokes emotions of unfairness, envy, revenge, superiority, and jealousy such that people want to reduce the unequal distribution.

- *Reciprocity norm*: A person's behavior is a function of another person's behavior in two or more periods. The reciprocity norm is the behavioral response of one player to another player when they confront each other over time. Like pro-social behavior, it represents a pattern of actions. A person's behavior can change in response to another person's actions. For instance, a person might be kind to a person who is also kind (positive reciprocity), mean to a person who is also mean (negative reciprocity), or alternate between the two (conditional reciprocity). A person might also selfishly help others, knowing that he or she may one day need someone else's help. Emotions such as fairness, revenge, fear, anger, and altruism can all be a part of reciprocity.

- *Trustworthiness*: A person trusts that some payoff-relevant information is true or that another person will take a certain action. For instance, a person may receive a message from another person signaling the optimal strategy to play, but the person who receives the message must decide whether to trust the other person and believe the message. When given a chance to opt out of a game and end it, a person must also consider trust when deciding whether to give another person a move in the game in the hope that this other person will do the optimal thing for both of them.

I have presented some imprecise definitions of emotions, categories of emotions, and the processes involved with emotions. This brief discussion of social preferences presents only a thin slice of the issues involved, but a detailed discussion of each emotion and how it relates to other emotions would overreach the scope of this book, which is to introduce basic game theory concepts. The interested reader should consult the references in this section for more precise definitions. I have included these concepts only to illustrate the notion of social preferences and distinguish terms when certain experimental results are discussed in later chapters.

4. EXAMPLE OF A SOCIAL UTILITY FUNCTION

Since social utility functions are such an important part of behavioral game theory, I will here briefly describe how these utility functions are constructed to provide an understanding of how these preferences are operationalized in a game theory model. Social utility functions rely on the common assumption that utilities are a direct function of a player's beliefs (i.e., the value of a utility depends critically on a player's beliefs). This assumption is referred to as "belief-dependent emotions." Elster (1998, 49) argues that certain emotions such as guilt, hatred, shame, regret, elation, fear, jealousy, disappointment, hope, and joy "are triggered by beliefs." According to this argument, what a person believes (motivated by emotions) can increase or decrease his or her utility (Geanakoplos, Pearce, and Stacchetti, 1989). To illustrate this idea, let b_i be a player i's beliefs about her own and an opponent's payoffs. We can model $u_i(b_i)$ as a player's payoff, or utility based on perceived beliefs. We will discuss this idea in more detail in later chapters, but for now let us just think about beliefs in terms of what our player knows—her own payoffs and the possible (or perceived) payoffs of her opponent. We can think of these beliefs as a discount factor that subtracts utility from a player's payoff or a bonus that increases a player's utility. For example, consider a player who gets a payoff of 3 for selecting an alternative, and let us assume that b_i can range from 0 to 1. A discount utility for 3 could be $u_i(b_i) = 3 \times (1 - b_i)$, so if $b_i = 0.50$, then 3 is discounted to 1.5. Similarly, if $u_i(b_i) = 3 \times (1 / b_i)$ and $b_i = 0.50$, then 3 increases to 6. To see how this assumption works, let us consider a player who has the choice of two actions:

1. Action 1: select a set of alternatives that grants our player a payoff of 3 and her opponent a payoff of 0.
2. Action 2: select a set of alternatives that grants both our player and her opponent a payoff of 2.

According to utility theory, our player should select Action 1 and receive the highest possible payoff of 3. However, our player may feel guilty about taking 3 and giving her opponent nothing. This feeling of guilt, which can be represented by a discount rate such that b_i diminishes the payoff of 3 to around 2 (i.e., $b_i = 0.334$), leads our player to select the fair outcome of (2, 2). We can also think of situations in which a player's utility increases as a result of feelings of superiority.

Rabin (1993) constructed a social utility function that models fairness and is particularly relevant here. Rabin's fairness utility functions allow players to derive utility from being nice to players who are perceived to be nice and being mean toward players who are perceived to be mean by allowing utility to increase or decrease based on notions of fairness. Let a_1 be a set of strategies for Player 1 and let $u_2(b_2, a_1)$ be the payoff that Player 1 believes that Player 2 will receive given her own choice of strategies and what she believes Player 2 will do. A fair or equitable payoff $u_F(b_i s)$ is defined as the average of what a player believes is the average of the high $u_H(b_i)$ and low $u_L(b_i)$ payoffs of another player. An index of fairness is $F = u_F(b_i) / (u_H(b_i) + u_L(b_i))$. To see how this works, assume that Player 1 perceives Player 2's payoffs as ranging from 0 to 100. Hence, $F = (50/(100 + 0))$ or ½, in which case payoffs higher than ½, such as 75, would be considered to be kind or would exceed fairness, and payoffs below ½, such as 25, would be considered to be mean. Hence, a social utility function that represents Player 1's kindness toward Player 2 would be $f(a_1, b_2) = (u_2(b_2, a_1) - F)$, where $u_2(b_2, a_1)$ is Player 2's payoffs given Player 1's strategy choice and Player 1's beliefs about Player 2's payoffs (or what Player 2 will do). Given this utility function, a person's utility can decrease if F is positive, increase if F is negative, or stay the same if $F = 0$. Thus, a choice function would have to allow for some sort of reciprocity behavior. That is, it would allow our player to be mean to players who are perceived to be mean (attempting to give them payoffs under 50) and to be nice to players who are perceived to be nice (attempting to give them payoffs higher than 50). This social utility function then captures emotions but stays within the framework of rationality.

I. SUMMARY

In this chapter, I attempted to show how the assumption of rationality allows for the definition of utility functions that map strategy selection to outcomes. Ordinal utility functions provide a ranking of alternatives that must obey the two ordering restrictions of connectivity and transitivity. Although these restrictions seem benign in certain situations (May's experiment and the three-person committee game), they can prohibit certain types of preference orderings. I also showed that utility functions can be represented in spatial terms when a player has an ideal point in one dimension and provided examples of how utility functions are induced in laboratory experiments. Finally, I discussed social preferences and the impact that emotions might have on behavior within a rational choice framework. In the next chapter, we will examine cardinal utility and expected utility theory, which place additional restrictions on players' preferences.

NOTES

1. Note that the symbol > means an individual strictly prefers one alternative over another and ≥ means greater than or indifferent to.
2. See McCulloch (1948).
3. Unfortunately, during the early days, rat experiments became associated with rational choice theory, thus leading to the unflattering label of "rat choice" theory. For other experiments that examine choice behavior of animals, see Basmann et al. (1975) and Green, Rachlin, and Hanson (1983). For pigeon research, see Heyman and Herrnstein (1986). For more rat research, see Kagel et al. (1975).
4. Computer simulations consider higher levels of dimensions, but this point is not addressed here.
5. To graph this function, you can place dots at each preference level and connect the dots from the left to the right.
6. There are notable exceptions to these alternatives. For example, some subjects might participate in a lottery when a cash or noncash prize is at stake, or subjects might be paid based on the correctness of their answers.
7. As a side note, early experiments that used spatial utility functions simply gave subjects a diagram exactly like Figure 3.8 and asked them to manually calculate the payoffs. Most experiments now use a computer program that calculates and displays payoff information to subjects.
8. Please note that this is not an exhaustive list. The study of emotions is a difficult area, since definitions of specific emotions are unclear and so-called "primary emotions" are combined with other emotions to form secondary emotions. Additionally, there are human feelings, which are a *result* of emotions. The classification used here is only for illustrative purposes.
9. The phrase "crowding out" means that some emotion or motivation replaces (or tries to replace) another emotion or motivation.
10. See Sugden 1982; Loewenstein, Thompson, and Bazerman 1989; Eisenberg and Mussen 1989; Rabin 1993; Elster 1998; Fehr and Schmidt 1999; Bolton and Ockenfels 2000; Fehr and Gächter 2002; Fehr and Fischbacher 2003; Xiao and Houser 2005.

EXPECTED UTILITY THEORY

A. EXPECTED UTILITY

1. EXPECTED VALUE AND SLOT MACHINES

Consider the following excerpt from a conversation between Charlie Rose, an interviewer for the television show *Sixty Minutes*, and Steve Wynn, the owner of a gambling casino in Las Vegas (CBS News 2009):

> "I want to understand a bit about the casino business," Rose remarked.
>
> "So do I," Wynn joked.
>
> He told Rose the only way to win in a casino is to own one, "unless you're very lucky."
>
> And he says, even when people are lucky, they usually gamble away their winnings.
>
> You have never known in your entire life a gambler who comes here and wins big and…walks away?" Rose asked.
>
> "Never," Wynn replied.
>
> You know nobody hardly that over the stretch of time is ahead?" Rose asked.
>
> "Nope," Wynn said.
>
> The customer's loss is Wynn's gain. He's a billionaire.

If people know that they are going to lose money, then why do they go to casinos to gamble? And if they win money gambling, why do they gamble it away? Our theory of preference choice thus far has only been concerned with the ordering and selection of alternatives. It has said nothing about risk, which is a relevant part of choice. Recall the example from the first chapter about whether a person in a sports car should pass a car that is pulling a trailer. I noted that assumptions about risk in the model could determine whether the driver of the sports car would pass the other car or not. In this chapter I will present the theory of expected utility, which allows us to model a player's risk-aversion. An answer to the question posed above is simply that some people derive utility from risky gambles. Although people might know the odds are against them, they are willing to spend money on a gamble because they think that they can beat the odds—that is, they have risk-loving preferences.

This chapter will focus on expected utility theory, which models risk behavior. Expected utility theory is concerned with preferences, where actions are defined as a lottery over possible outcomes and this lottery can be measured and assigned a value. Imagine that the amount you have on an ATM card is your value. An ATM machine will only allow you to withdraw the amount of cash that is on the card, which represents the cash value for that machine. Expected value is like the cash value of a slot machine at a casino, or what a person expects to win from a particular slot machine. We assume that a person prefers those slot machines that have the highest possible payout and offer the greatest probability of winning. All the slot machines in the casino have an expected value, and people always prefer the slot machines that have the highest expected value. In fact, a person always prefers slot machines with higher expected values, even when those values are so high that they cannot be adequately measured.

2. THE ST. PETERSBURG PARADOX

Is the assumption that people always prefer higher expected values reasonable? Consider two people who have just met at a gambling establishment,[1] Olga and Igor. Although they have just met, Olga doesn't think that Igor is very bright, so she decides to try to win his money by offering him the following bet. She shows Igor a penny and tells him it is not weighted and Igor checks to make sure. Olga explains that she will flip the coin and if it comes up heads, she will pay him two dollars and the bet will end. She further explains that if the coin comes up tails, she will flip it again and pay him twice the amount, and that she will continue doing so each time as long as she keeps flipping tails. Igor doesn't understand, so Olga tells him that she will pay him 2^n dollars, where n is the number of consecutive times a tail is revealed. She then explains that in the first round the probability of a heads or tails is 0.50, so his expected value is one dollar (0.5×2). If a tail appears, his expected value will be 2^1 in the first round, or two dollars; if another tail appears in the second round, his expected value will be 2^2, or four dollars; if a tail appears in the third round, his expected value will be 2^3, or eight dollars. If consecutive tails continue for 10 rounds, then his expected value will be 2^{10}, or 1,024 dollars, and in 20 rounds, his expected value will be in the trillions of dollars. After explaining all of this, Olga asks Igor if he is willing to give her a million dollars to play this bet. Igor balks. But Olga tells him that because the expected value is in the trillions (or really infinite), one million dollars is a small fee to pay to participate in this bet. Igor scratches his head and wonders aloud, "Now how many tails in a row do I need?"

This example is the St. Petersburg paradox suggested by Bernoulli (1738), which illustrates the importance of expected utility theory and the need to place bounds on behavior.[2] The paradox concerns the problem of assuming a linear relationship like the one illustrated in Figure 3.1. This assumption of a linear relationship treats the expected value the same for everyone, so that everyone "equally" always prefers more to less of a value (i.e., everyone values and enjoys 100 dollars the same). The problem that Bernoulli realized was that value in terms of utility is subjective and differs in importance from individual to individual. For example, 100 dollars is worth a lot more to a homeless person than to a billionaire. Bernoulli's suggestion was the notion of diminishing marginal utility, in which bending the linear utility function to create a concave utility function allows for differences in behavior. Diminishing marginal utility means that for low values of particular things, people derive large utility, but for higher values of the same things, they derive only marginal or slight utility. In fact, at a certain point when the utility line is flat, people are indifferent to higher values. Hence, by bending the straight utility function line (or by considering a differently shaped utility function), we can discuss behavior that is more realistic. In lay terms, bending lines, adding more dimensions to create more lines to bend, adding parameters to the lines, and then interpreting behavior from the resulting creation are, in a nutshell, what utility theory is about.

In the next sections, I will discuss expected utility theory and explain how risk behavior is modeled. I will then outline an alternative to the standard model of risk, prospect and regret theory. In the fifth and sixth sections, I will examine anomalies of expected utility theory. In the seventh section, I will discuss alternatives to expected utility theory. In the last section, I will show how preferences defined as lotteries are implemented in laboratory experiments by illustrating a binary lottery experiment.

B. EXPECTED UTILITY THEORY

1. USING CARDINAL VALUES IN A UTILITY FUNCTION

Recall that when ordinal utility is assumed, the intervals between values have no meaning, so that if A is preferred to B, no claims can be made about *how much* A is preferred to B. When cardinal utility is considered, the intervals between values gain meaning. For example, imagine a scale that ranges from 0 to 100 on which a player's preference for Alternative x is assigned a cardinal utility of 90 and his preference for Alternative y is assigned a cardinal utility of 89. Drawing a comparison between the two alternatives, we can make the statement, "Alternative x is slightly more preferred than Alternative y." If a player's preference for Alternative x is assigned the same cardinal utility of 90 but his preference for Alternative y is assigned a cardinal utility of 9, assuming the two scales are equivalent, we can make the statement, "Alternative x is much more preferred than Alternative y." These are statements you cannot make about ordinal utility. Cardinal utility provides a method to measure intensity of preferences among various alternatives. Although interpersonal comparisons of utility (that is, comparing intensity of preferences for alternatives among players) can be made using cardinal utility, these comparisons are problematic (see Abrams 1980 and Problems 4.3 and 4.4).

Since cardinal utility allows us to measure the intensity of preferences, it also allows us to measure the risk that is involved in decision-making. That is, a person who only slightly prefers one alternative to another may behave differently than a person who greatly prefers one alternative to another. A person who greatly prefers one alternative to another may be willing to take a greater risk to achieve that alternative, as opposed to a person who may not be willing to exert any risk at all, since she only slightly prefers one alternative to another. Consider a choice between A and B and imagine that, given the same decision-making environment and constraints, a risk-loving person might prefer A and a risk-averse person might prefer B. As we will see, by varying the risk behavior of the two players, we can vary their preferences for alternatives.

To understand this situation, let us define a lottery as a choice between two or more risky alternatives, such that a player's preferences are defined over lotteries. A player's preference is determined by adding the value of the prize (the cardinal utility value or what you would get if you won the lottery) to the probability of winning the prize for all possible values. This number allows the player to rank the value of a prize according to the worth of the prize and the probability of winning it. A player prefers alternatives that have a higher probability of winning as well as a higher value. To define this a little more formally, let p be some probability ranging from 0 to 1. So, if $p = 0.3$, then $1 - p$ must equal 0.7. A lottery (L) involving x and y is represented as $L = px + (1 - p) y$. We can plug our cardinal utility values into this lottery and generate expected utilities (i.e., what a player expects to win given the probabilities associated with each cardinal utility value).

Von Neumann and Morgenstern (1944) formally proved the expected utility theorem that allows us to define utility functions that use cardinal values. In their honor, a utility function that involves lotteries is referred to as a von Neumann-Morgenstern (vNM) utility function.

2. PREFERENCES OVER LOTTERIES VS. PREFERENCES OVER OUTCOMES

As a point of clarification, in the last chapter's discussion of ordinal utility functions, players had preferences over outcomes, so that, for example, a person had utility for 10 dollars and the money was the object of value. vNM utility functions suggest that people have preferences over lotteries and it is the lotteries that are valued, not the actual outcomes. For example, assume two lotteries:

> Lottery A (p): A person has a 0.50 chance of winning 100 dollars and a 0.50 chance of winning nothing.
> Lottery B (q): A person has a 0.10 chance of winning 100 dollars and a 0.90 chance of winning nothing.

A vNM utility function assumes a person has preferences over the distribution $p = (0.5, 0.5)$ and $q = (0.10, 0.90)$, such that over the set of outcomes $x = (\$100, \$0)$, the distribution $(0.5, 0.5)$ is preferred to the distribution $(0.10, 0.90)$. In this case, preferences are defined over these distributions (i.e., probability numbers) and not over the outcomes (i.e., \$100, \$0). Defining preferences over distributions allows us to deduce the implied preferences over the underlying outcomes.

You may ask yourself, "Why do we want to define preferences over lotteries?" Ordinal utility functions are used when players know the state of nature, which allows them to determine the consequences of their actions before making decisions. However, in many instances, an individual does not know with certainty the state of nature that does or will exist, and defining preferences over lotteries allows us to model this uncertainty. To understand this idea, let us go back to Olga and Igor. After Igor refuses her bet, Olga thinks that he might be smarter than he looks, and for some reason, although she isn't quite sure why, she becomes attracted to him. Olga wants to see if Igor likes her back, so she considers throwing a party and inviting him to meet some of her friends. Olga conjectures that if Igor has a really good time at the party (dancing and conversing with her friends), then this will indicate that he likes her, but if he has a bad time at the party (behaving like a wall-flower), then this will indicate that he does not like her. So Olga must decide whether to hold her party, despite the uncertainty that she only wants to have a party if she thinks Igor is going to have a good time but does not want to throw a party if she thinks he is going to have a bad time. As a result, Olga cannot observe two states of nature prior to her decision: one in which Igor enjoys the party and one in which Igor hates the party. This situation illustrates decision-making under uncertainty. Olga cannot be certain whether Igor will like the party or not before she has to plan the party. Notice how risk is involved in this decision. If Olga is a risk-loving person, she is more likely to throw a party, but if she is risk-averse, she is more unlikely to throw a party.

3. FURTHER RESTRICTIONS ON CHOICE

To define expected utility theory, we need to make three additional assumptions about preferences. I should point out that there is no standard way in which these terms are defined and different scholars use different terminology to represent the same assumption. Therefore, the terms that I use may be different from the terms that other scholars use. In the following definitions, assume that x, y, and z are three lotteries.

> *Independence*: If an individual prefers x to y, then px + (1− p)z > py + (1− p)z, and the individual should also prefer x to y when a third lottery, z, is added to each x and y term. Hence, if x > y, then x + z > y + z.

This assumption means that an individual's selection between two lotteries x and y should be unaffected by a third and irrelevant lottery such as z. In other words, if two lotteries are mixed with a third lottery, the preference ordering of the first two lotteries should not be affected by the third lottery. This assumption ensures that preferences are well behaved and prevents preference reversals from occurring (e.g., a true preference A > B can be shown to be B > A).[3]

> *Continuity*: If x > y > z, when px + (1− p)z, there is a p that makes y equally preferred to this lottery.

This assumption means that given the possible combinations of x and z, there should be a value of p for which the individual will be indifferent between this mix and the lottery y (i.e., a player will be indifferent between his best and worse outcomes). Continuity places a bound on outcomes that excludes infinite utilities for outcomes that can create unpredictable behavior (i.e., the preference relation over the lotteries is closed) so that outcomes can be related to a numerical scale. In other words, if a person gets unlimited positive or negative utility from an alternative, then it may not be possible to find a lottery for which a player is indifferent between a best and a worst outcome—that is, if you cannot find the top or bottom of a line, then it is impossible to find the middle of the line. Therefore, this assumption assures that a top outcome and a bottom outcome exist that can be measured. Continuity implies that small changes in probabilities do not affect the ordering of the preferred lotteries (e.g., if a trip on a train is preferred to a trip in a car, then a trip on a train with a small chance of derailment is still preferred to a car trip).

Reduction of compound lotteries: A compound lottery can be reduced to a simple lottery with the same expected value, and an individual should be indifferent between the two lotteries.

An example of a simple lottery is paying money to roll a pair of dice and winning some money (presumably more than you paid to participate in the lottery) if you roll a 7 or an 11. A compound lottery is when you participate in a lottery to win a ticket to participate in another lottery (or lotteries) for some value (money). For example, in a two-stage compound lottery, you have to roll a 7 or an 11 in the first stage, and if you succeed, you get to roll again in the second stage and win some money if you roll a 7 or an 11 again. This assumption means that a person should be indifferent between a simple lottery in which he has a 25 percent chance of winning 100 dollars and a two-stage lottery in which he has a 50 percent chance of winning 100 dollars in each stage or a 25 percent chance of winning 100 dollars overall. In both types of lotteries, the expected value is the same (25%), so a person should be indifferent to either lottery. Repeated games are often modeled as compound lotteries, and the derivation of equilibrium depends on this assumption of reduction of compound lotteries. This assumption is important, because it means that our players only care about the distributions of the lotteries, which allows us to focus on simple lotteries. In short, reduction of compound lotteries means that a player does not care about the presentation of the lottery (whether it is simple or complex), only about the final probability calculated from the lottery (assuming that each lottery is independent). That is, it does not matter if the slot machine is a plain black box or brightly decorated with bells, lights, and whistles. As long as both slot machines produce the same outcome, their appearance does not matter.

Although these assumptions may appear technical, they simply mean that we can define an expected utility function by modeling preferences over simple lotteries. This utility function is linear for probabilities represented by straight, parallel indifference curves, and it ensures that increases in probability increase the chance of winning the lottery, so that players prefer higher chance lotteries to lower chance lotteries. This idea is von Neumann and Morgenstern's (1944) famous expected utility theorem, which proves that if these conditions are fulfilled, then a player's preferences over lotteries can be represented by cardinal utility. Morrow (1994, 32) notes:

> The Expected Utility Theorem provides us with a way to represent deciders' preferences on a cardinal scale. The differences in utility among outcomes allow us to judge what risks deciders will accept. Because utilities are calculated from deciders' willingness to take risks, they measure relative preference among outcomes by the probabilities of obtaining different outcomes.

4. CALCULATING EXPECTED UTILITY

Let us consider an American football example. In this game, an offense attempts to move a ball forward across a goal and a defense attempts to prevent the forward movement of the ball. The offense can either run or throw the ball, and the defense can use either a run defense or a pass defense. Since teams plan their strategies in a huddle, neither team knows what strategy the other will select, uncertainty exists, and this becomes a game of probabilities. If both teams are equal in strength, then the winner is determined by the flip of a coin. However, both teams are usually not equal, and one team will often have a better run offense and the other team will often have a better pass defense. As a result, a team with a better run offense uses that strategy more often and a team with a better pass defense uses that strategy more often. During gameplay, when the offense runs the ball and the defense plays a run defense, the offense will have an advantage, but when the offense runs a pass play and the defense plays a pass defense, the defense will have an advantage. Both teams mix up strategies, using some more than others, to outguess their opponent. The game is interesting because it has many equilibria.

This section does not aim to show how to solve equilibrium for football games but to learn how to calculate expected utility. Assume that the offense wants to run 60 percent of the time and pass the ball 40 percent of the time, whereas the defense wants to play a run defense 50 percent of the time and play a pass defense 50 percent of the time. The distributions of 60/40 and 50/50 are what we are referring to as lotteries. Expected utility is the number that represents the expected value a coach would expect to receive if he plays the strategies with the specified lotteries. In this example, utility is defined for the offense in terms of yards gained and for the defense in terms of yards minimized. Assume that the offense devises a plan to pass the football 40 percent of the time, so that over the course of the game it will gain 500 yards, and run the ball 60 percent of the time to gain 400 yards. The expected utility for the offense (EU_O) from playing 40 percent pass and 60 percent run can be written as

$$EU_O(P, R) = 0.40(500 \text{ when passes}) + 0.60(400 \text{ when runs})$$

The expected utility for this lottery is

$$EU_O(P, R) = 0.40(500) + 0.60(400) = 440$$

An alternative game plan the offense can use is to pass the ball 30 percent of the time, expecting to gain 500 yards, and run the ball 70 percent of the time, expecting to gain 400 yards. The expected utility for this second game plan, or EU_{02}, is

$$EU_{02}(P, R) = 0.30(500) + 0.70(400) = 430$$

We assume that a player prefers lotteries with higher expected utility, so the offense would prefer to use the first game plan, since $EU_O(P, R) > EU_{02}(P, R)$, or 440 > 430. Note that this is not a game, but only what the offense expects to gain if all plays are perfectly run and they can figure out what the defense is going to do every play. Hence, this calculation does not account for offensive or defensive execution of specified plays against each other.

The general form for expected utility where Q is the utility and p is a probability is

$$EU(Q) = p_1(Q_1) + p_2(Q_2) \ldots p_n(Q_n)$$

The sum of all ps should equal 1.

This expected utility function does not assume risky behavior, which is a relevant part of decision-making. Risk in our football example would be when the offense has one last play and the coach must decide between kicking a field goal and settling for three points or playing for the six points of a touchdown, knowing that if the play fails, the team will get no points. The time at which this decision is made during the game produces different degrees of risk, but in general, a coach who has risk-loving preferences would be more likely to try for seven points, while a coach who has risk-averse preferences would more likely settle for three.

C. MODELING RISK

1. WHAT IS RISK?

Since our utility function concerns preferences over lotteries, we are able to measure risk in terms of preferences. By risk we mean how much of a gamble a person is willing to take, which differs among individuals. You may know people who love to climb mountains, sky-dive, wrestle alligators, swim with sharks, or hang out with grizzly bears. You also probably know people who would be frightened to do these things and would rather just cuddle up under a cozy blanket and read stories about such dangerous adventures. In terms of risk, those people who actually engage in these activities are considered to be risk-acceptance or risk-loving people, while those people who do not are considered to be risk-averse or risk-hating individuals. Mathematically, risk can be modeled by assuming different shapes of a player's utility function. Risk-neutral behavior is represented by a linear utility function $f(x) = a + bx$ (see Figure 3.1), representing neither risk-averse nor risk-acceptance behavior.

2. MODELING RISK-AVERSE VS. RISK-ACCEPTANCE BEHAVIOR

To examine risk, we must rescale the monetary dimension (our cardinal values) using non-linear scaling to bend this straight line. Risk-averse behavior can be modeled by using a concave upward sloping utility function in which monetary payoffs are rescaled using a square root function $f(x) = \sqrt{x}$. Risk-acceptance behavior can be modeled by using a convex utility function in which monetary payoffs are rescaled using an exponential function $f(x) = 2^x$. To understand these ideas, let us consider three different types of people: risk-neutral people, risk-averse people, and risk-acceptance people. All of these types are given the following choice: they can either get 50 dollars with certainty or they can participate in a lottery in which they have a probability p of winning 100 dollars or a probability 1– p of winning nothing (we assume p = 0.5). Regardless of type, each person has the same utility for the worst outcome of 0 dollars and the best outcome of 100 dollars (because these values are at the endpoints of the utility function), and so the utility for this lottery will remain the same, $\frac{1}{2} u(0) + \frac{1}{2} u(100)$, for all types. However, the utility that people have for the middle prize of 50 dollars varies, since this value will move as the utility function changes shape (as we bend the line). Let us first consider the type of people who are risk-neutral, depicted in Figure 4.1.

This linear utility function represents the case in which a person is indifferent between getting 50 dollars with certainty and participating in a lottery between 0 dollars and 100 dollars ($\frac{1}{2} u(0) + \frac{1}{2} u(100)$). In this graph, the utility for the lottery is equal to the utility of 50 dollars. That is, $\frac{1}{2} u(0) + \frac{1}{2} u(100) = u(50)$.

Now let us consider risk-acceptance people, represented by a concave upward utility function, as displayed in Figure 4.2.

Notice in the figure that because of the shape of the utility function, the utility for 50 dollars has shifted downward on the axis so that it is now below the utility for the lottery, ½ $u(0)$ + ½ $u(100)$. Therefore, these types of people would prefer to take the lottery than to get 50 dollars with certainty. That is, ½ $u(0)$ + ½ $u(100)$ > $u(50)$.

Finally, consider risk-averse people, represented by a convex utility function, as depicted in Figure 4.3.

Since the shape of the utility function is upward sloping, the utility for 50 dollars is now higher than the utility for the gamble, ½ $u(0)$ + ½ $u(100)$. These people are risk-averse

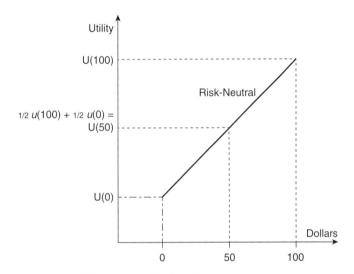

FIGURE 4.1 Risk-neutral utility function

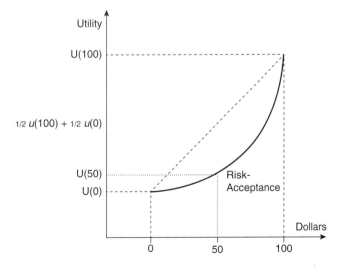

FIGURE 4.2 Risk-acceptance represented by a concave upward utility function

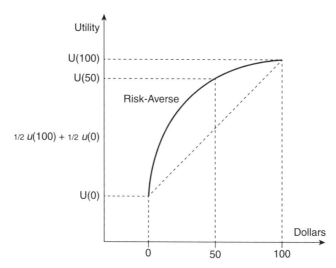

FIGURE 4.3   Risk-averse utility function represented by a convex utility function

Source: Reprinted with permission of University of Chicago Press; Hogarth and Reder 1986.

because they would prefer to get 50 dollars with certainty than to have the opportunity to participate in the lottery. That is, $u(50) > \frac{1}{2}\, u(0) + \frac{1}{2}\, u(100)$.

Consequently, depending on the shape of the utility function, we can model risk in two different ways. One problem is that these two ways do not capture a lot of behavior that is observed in terms of risk. For instance, on November 11, 2011, so many people wanted to play 11/11/11 in state lotteries that lottery officials capped the number of tickets that could be sold with this number.[4] It was estimated that lottery sales on this day increased by 5 to 10 percent. In this situation, the lottery officials deliberately reduced risk in a non risky situation and people bought more lottery tickets as the odds of winning decreased. In our model of risk, people cannot behave in this manner because of the assumptions of convexity or concavity.

D. FRAMING EFFECTS AND ALTERNATIVE THEORIES OF RISK

1. FRAMING

Framing is another important aspect of behavioral game theory. By framing I mean that the manner in which a problem is posed can impact its outcome. For example, Ariely (2008) conducted an experiment in which the same coffee was served to people in a Styrofoam cup and a fancy coffee mug, and most people thought the coffee tasted better in the fancy mug. This example encapsulates the problem with framing: same coffee, different cups, different outcomes. For instance, say that you conduct an experiment in the middle of a busy casino and then conduct the same experiment in a sterile computer laboratory. Will you get the same results from both trials? Maybe,

maybe not. Or say that you made a decision at 1:00 A.M. and then made the same decision at 1:00 P.M. Would the decisions be different? It would depend on the nature of the decision.

Framing is a constant concern because everything has a frame, so the problem is identifying which frames bias outcomes and which frames do not. As Chapter 2 noted, experiments that examine game theory models use sterile environments to mitigate any possible framing effects. However, if everything has a frame, then the sterile environment itself becomes a frame whose impact on behavior needs to be understood. It is important to understand frames to understand how decisions are made. The effect of frames on decision-making is studied in decision environments such as public opinion responses (see Druckman 2001; Chong and Druckman 2007).

2. PROSPECT THEORY

Framing also impacts risk behavior. Kahneman and Tversky (1979) argue that subjects in experiments view losses and gains differently in terms of risk because of framing effects. They note that "framing is controlled by the manner in which the choice problem is presented as well as by norms, habits, and expectancies of the decision maker" (Tversky and Kahneman 1986b, 273). Hence, the environment in which the choice is made has an influence on which choice is made. They (1986b, 274) present the following experiment, in which subjects are asked to choose among various hypothetical lotteries (or prospects):

Problem 1: Assume yourself richer by $300 than you are today. You have to choose between
    a. A sure gain of $100
    b. 50 percent chance to lose nothing and 50 percent chance to gain nothing

Problem 2: Assume yourself richer by $500 than you are today. You have to choose between
    a. A sure loss of $100
    b. 50 percent chance to lose nothing and 50 percent chance to lose $200

The expected value is the same in both problems, so if expected utility theory is correct, subjects should respond to a and b in both problems with a 50/50 distribution. But in Problem 1, most subjects selected a (72%), which indicates risk-aversion, and for Problem 2, most selected b (64%), which indicates risk-acceptance. Tversky and Kahneman argue that although the expected values were the same, the context or the frame for the problem differed. In Problem 1, the phrase "a sure gain" triggers a risk-averse response, whereas Problem 2's phrase "a sure loss" triggers a risk-acceptance response. Consequently, Tversky and Kahneman argue that a utility function should be both risk-averse and risk-acceptance. They propose a different method to model risk, illustrated in Figure 4.4.

This model includes a reference point or status quo point. As a person gains utility or money relative to her status quo, the shape of the utility function changes. The function is concave above the status point and convex below the status quo. Its shape illustrates a case in which people like to gain money but really hate to lose it.[5] The reference point thus provides a frame that pulls decisions.

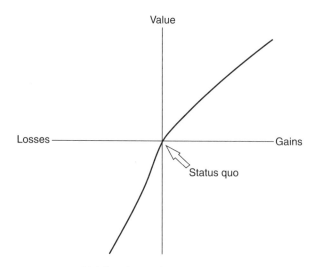

FIGURE 4.4   Risk function under prospect theory

3. REGRET THEORY

Another alternative theory of risk is regret theory (Loomes and Sugden 1982; Bell 1982). This theory argues that when people make decisions while confronted with uncertainty, they often experience regret after learning about the consequences of their decisions. The degree of regret differs depending on the context of the loss. Loomes and Sugden give the example of losing 100 dollars. Under expected utility theory, a loss is a loss—if a person loses 100 dollars, there is no consideration of the context in which the money is lost. However, Loomes and Sugden point out that losses in different environments can, like framing effects, affect the expected utility, so that a loss of 100 dollars in one decision environment can be greater than a loss of 100 dollars in a different decision environment. They point to the case in which a person's taxes are raised 100 dollars. In this instance, a person may not feel regret over losing 100 dollars, since she may believe that because taxes are set by law, there is nothing she can do but accept her loss. However, if a person bets on a horse that loses the race, added regret must be considered. The person feels horrible, since either she should have not placed the bet or she should have placed the bet on another horse. Under regret theory, then, a 100 dollar loss is treated differently depending on the circumstances. Here, utility in the second case should be diminished more than in the first case. The purpose of this theory, like prospect theory, is to show that people can behave in a risky way (i.e., have convex utility functions) and at the same time exhibit adverse behavior (or have concave utility functions). Behavior in this theory is driven not by the desire to maximize utility, but by the desire to avoid post-decision regret.

While most would agree that both prospect theory and regret theory are good ways to measure risk, their application in the game theory literature has been sparse. One reason is that assuming concavity or convexity is a rather straightforward process that is easy to implement in standard models. However, assuming both of them together in a utility function is complicated in standard models. A notable exception is quasi-concave utility functions, which are used in spatial models of elections and committees (see Ordeshook 1986).

E. ANOMALIES TO EXPECTED UTILITY THEORY

There are many anomalies to expected utility theory, or behavior that contradicts the theory's assumptions. The literature on this subject is so vast that in this section I will only present a few examples.

1. THE ELLSBERG PARADOX

Let us again consider Olga, who has decided to pass on her party plans and instead go to her favorite gambling establishment.[6] Once there, Olga is led to a private room, where an employee of the club shows her two opaque urns that look exactly the same. She is told that inside each urn is a marking that identifies it as the left or the right urn, and that the left urn contains 50 red marbles and 50 blue marbles (the distribution is known), and the right urn contains 100 red and blue marbles (the exact distribution is not known). The employee tells her that she must select a marble from one of the urns and that she will win 100 dollars if she correctly guesses the urn from which it came. Olga agrees to the gamble and selects a red marble. The employee asks Olga whether the marble came from the left or the right urn. Conjecturing that her chances of being right are better with the left urn, which has a known distribution of red and blue marbles, Olga selects the left urn. This selection also implies that she thinks there are more blue marbles than red marbles in the right urn. Olga wins 100 dollars and the employee asks her to play again for double or nothing. Confident, Olga agrees and selects a blue marble. When the employee asks her from which urn the marble came, Olga says the left urn, loses the bet, and owes the gambling establishment 200 dollars. After paying her debt, Olga is left kicking herself, because she understands that based on her previous selection, she should have chosen the righthand urn, since she perceived it (based on her expected utility calculations) to have more blue marbles.

The problem in this example is that, like Olga, most people avoid the urn with the unknown distribution. So it is something about uncertainty that causes violations of expected utility theory. People seem to prefer certain information to uncertain information. Some scholars have argued that this preference is fear of the unknown or what is referred to as ambiguity aversion (shying away from a gamble in the absence of knowledge about the probabilities). That is, players' behavior tends to sway from the prescribed expected utility behavior when players do not know probabilities or confront ambiguous decision-making situations (see Knight 1921; Fox and Tversky 1995). In the urn example, the ambiguous decision is simply made by selecting the urn with the known probability.

2. FRAMING AND REFERENCE POINTS

After losing at the gambling establishment, Olga checks her wallet and finds that she only has two 10-dollar bills left. To help her depression, she decides to go to a show. When she gets to the show, she discovers that she has lost 10 dollars. Olga still has one 10-dollar bill left in her wallet, so she has to decide if she will buy a ticket to the show or not, given that she has just lost 10 dollars. Consider another similar situation. Olga buys a ticket to the show in advance and after losing at the gambling establishment checks her wallet and sees that she has her 10-dollar ticket to the 1:00 show and a single 10-dollar bill. Since it is almost 1:00, she rushes to the show, but when she gets there, she realizes that she has lost her ticket but still has 10 dollars in her wallet—enough to buy another ticket. Should Olga buy another ticket to the show?

In terms of expected utility theory, these two situations produce the same expected value of 10, but in the first situation, Olga decides to buy a ticket with her last 10 dollars, while in the second situation, she decides that she does not want to spend her last 10 dollars to replace her lost ticket.[7] Tversky and Kahneman (1986a, 135) comment:

> The marked difference between [these two cases] is an effect of psychological accounting. We propose that the purchase of a new ticket in [the second case] is entered in the account that was set up by the purchase of the original ticket. In terms of this account, the expense required to see the show is $20, a cost which many of our respondents apparently found excessive. In [the first case], on the other hand, the loss of $10 is not linked specifically to the ticket purchase and its effect on the decision is accordingly slight.

In another dilemma, Tversky and Kahneman show how expected utility can be biased as a result of reference points or framing.

> *Case 1*: Suppose that Igor just won 30 dollars and is asked to enter Lottery 1.
> *Lottery 1*: Igor is first given a choice between flipping a coin and not flipping it. If Igor flips the coin and it comes up heads, he wins nine dollars, and if it comes up tails, he loses nine dollars.
> *Case 2*: Suppose that Igor has 0 dollars and is asked to enter Lottery 2.
> *Lottery 2*: Igor is first given a choice of flipping a coin or getting 30 dollars with certainty. If Igor flips the coin and it comes up heads, he wins 39 dollars. If he flips the coin and it comes up tails, he wins 21 dollars.

In Case 1, Igor decides to flip the coin, and in Case 2, he decides not to flip the coin. Igor's behavior is confirmed by Tversky and Kahneman's experiments showing that 70 percent of subjects chose to flip a coin in Case 1 and 43 percent chose to flip a coin in Case 2. Again, both cases have the same expected value, but choice varies. These experiments show that subjects' choices were biased by the reference point, or by whether or not a subject started the lottery with money or did not start the lottery with money.

3. TIME INCONSISTENCY

Another anomaly that occurs when considering expected utility concerns time, which was mentioned in the last chapter. As we have discussed thus far, expected utility is time specific—its assumptions only hold for a slice of time or at the moment a decision is being made. That is, behavior is static (we will examine dynamic behavior in Chapter 11). Nevertheless, expected utility theory can be impacted by time. There is a difference between the value of long-term expected payoffs and the value of short-term expected payoffs. For example, Olga smokes cigarettes and knows that in the long run, this decision may damage her health. However, because she believes these consequences will occur in the distant future, she takes the risk and smokes anyway. However, if Olga believed that by smoking cigarettes she would die within a week, then she would probably quit smoking. Thus, payoffs, or the utility that people receive, can also be viewed in a long versus short timeframe. Ainslie and Haslam (1992) conducted an experiment in which they offered subjects a choice between 10 dollars on the day of the experiment or 11 dollars in one week. Naturally, most subjects chose to take the 10 dollars on the day of the experiment. Subjects were then offered 10 dollars a year from the day of the experiment or 11 dollars a year and a week from the day of the experiment. Under these circumstances, most preferred to wait an additional week and get 11 dollars. Hence, in a short timeframe $u(10)$ > $u(11)$, but in a long time frame $u(11)$ > $u(10)$. This finding violates the independence assumption of expected utility theory.

This time inconsistency can be overcome if we consider how people discount the future. We can add a discount rate to a utility function to adjust for this inconsistency. For example, consider the utility for 100 dollars. Say that a person who is not concerned with the future (in terms of getting his payoff) has a discount rate of 0.9. This would mean that in the future his utility would be worth $0.9 \times u(100)$, or 90. However, an individual who is concerned with the future (and wants his payoff sooner rather than later) might have a discount rate of 0.1. In terms of utility, his future payoff would be worth $0.1 \times u(100)$, or 10. These types of discount rates are used in repeated games to attempt to model the element of time (see Morrow 1994).[8]

F. ALTERNATIVE THEORIES TO EXPECTED UTILITY THEORY

1. BOUNDED RATIONALITY

These inconsistencies in expected utility theory motivated psychologists to study rationality. Hogarth and Reder comment (1986, 5): "The challenge to the psychologist is to construct a theory of the process that leads to preference reversals and to see whether this might generalize to domains other than choices between gambles." One such theory is called bounded rationality and is discussed by Simon (1965; 1985). Bounded rationality means that people do not actually optimize decisions but simply do the best that they can given the decision problem they confront. Simon (1985, 295) notes:

> [Bounded rationality] . . . describe[s] a person who is limited in computational capacity, and who searches very selectively through large realms of possibilities in order to discover what alternatives of actions are available, and what the consequences of each of these alternatives are. The search is incomplete, often inadequate, based on uncertain information and partial ignorance, and usually terminated with the discovery of satisfactory, not optimal, courses of action.

This type of behavior is referred to as "satisficing" behavior and describes a pattern of behavior in which an individual might make mistakes now and then but when a decision has to be made uses a simple heuristic. For example, in a presidential election, a voter may not understand all of the issues that the two candidates are espousing, but he can discern that one candidate is liberal and the other conservative, so he makes a decision based on this heuristic, rather than on the specific issues for which the candidates are advocating.

2. THE BPC MODEL

Gintis (2009) proposes a different version of expected utility theory that relies on fewer assumptions than the traditional theory. He refers to his model as the BPC model, with BPC standing for beliefs, preferences, and constraints. Gintis argues that rationality only requires assumptions about preference consistency that do not require supercomputing abilities or perfect knowledge. The BPC model only relies on three assumptions about preferences: completeness, transitivity, and independence. In his model, people do not have to behave selfishly or out of self-interest, since this type of behavior does not violate his notion of preference consistency. He notes (2009, 7):

> Decision theory does not presuppose that the choices people make are welfare-improving. In fact, people are often slaves to such passions as smoking cigarettes, eating junk food, and engaging in unsafe sex. These behaviors in no way violate preference consistency.
>
> If humans fail to behave as prescribed by decision theory, we need not conclude they are irrational. In fact, they may simply be ignorant or misinformed. However, if human subjects consistently make intransitive choice over lotteries, then either they do not satisfy the axioms of expected utility theory or they do not know how to evaluate lotteries. The latter is often called *performance error*. Performance error can be reduced or eliminated by formal instruction, so that the experts that society relies upon to make efficient decisions may behave quite rationally even in cases where the average individual violates preference consistency.

What these two theories have in common is the assumption that people are not perfect expected utility maximizers; on the contrary, people often make mistakes and use heuristics to make decisions.[9]

G. BINARY LOTTERY EXPERIMENTS

Roth and Malouf (1979) conducted experiments in which subjects had preferences over lotteries. In these experiments, two players were given an expected prize that they could win. For example, one player might be assigned a prize of five dollars and another player might be assigned a prize of 20 dollars. Subjects then bargained over how to distribute lottery tickets that determined their probability of winning their prize. If one subject were able to bargain and get 60 percent of the tickets, then she would have a 60 percent chance of winning her prize. Subjects in this experiment were seated at computer terminals so that their identities were anonymous, but they were able to send text messages to other subjects and make various proposals. For example, one subject might propose a 60/40 split in which he would get 60 percent of the tickets and his opponent would get 40 percent. Subjects were placed on a timer and if no agreement was reached, then each subject received nothing. These games are referred to as binary lottery games, since players either receive a particular allocation of lottery tickets or receive nothing. The subjects' utility functions in these games are normalized between 0, where there is no payoff, and 1, where the player wins the lottery. This allows the subject's utility from any lottery to range between 1 and 0, which represents the probability of winning the lottery. This neat experimental trick allows a researcher to consider a model that assumes strategies as lotteries but can treat these strategies like pure strategies in the analysis.

Nash's (1950) bargaining model motivated this example, since it assumed that the solution should not be sensitive to transformations of payoffs. That is, the solution to the bargaining game should not depend on whether or not players knew each others' prizes. Roth and Malouf (1979) varied the information that players had about the other prizes. In one trial the subjects did not know their opponent's prize, and in another trial the subjects did know their opponent's prize. Contrary to the theory, there were differences in outcomes. When subjects did not know another subject's prize, an equal split was generally proposed, and when they did know another subject's prize, the offers deviated from this even-split outcome, with one subject demanding more of the tickets.

This experiment thus offers evidence supporting the notion of inequality-averse social preferences discussed in the last chapter. Rationality assumes that players only care about their own winnings, with the earnings of other subjects being irrelevant. However, these experiments show that people do care about their winnings in relation to other people's winnings. Hence, subjects exhibit the presence of a social utility function in that they care about money, but they also care about issues such as fairness and equality.

H. SUMMARY

In this chapter, I have shown how cardinal utility values are used in preferences to formulate expected utilities. In terms of payoffs, these types of preferences reveal what a player expects to gain from certain lotteries. I illustrated how mathematical functions such as convex and concave functions can be used to model risk-aversion and presented two alternative representations of risk that do not use the classic assumption, prospect and regret theory. I discussed a few anomalies to expected utility theory in which choice behavior is altered as a result of the environment in which a player is placed. I presented some alternative paradigms to expected utility that argue that people's tendency to make mistakes should be reflected in the models. Finally, I presented the details of several experiments, including a binary lottery experiment, which allowed for preferences that replicate vMN utility functions. These experiments disproved the conjecture that knowledge about other players' payoffs should not impact equilibrium predictions.

NOTES

1. This is just a made-up social hang-out place where people go to make bets with other people.
2. This example is referred to as the St. Petersburg Paradox because Bernoulli first published it in the *St. Petersburg Academy Proceedings* in 1738. (The first English translation was done in 1954.)
3. This assumption also ensures that indifferent curves are linear and parallel.
4. See http://www.lottery.com/stories/lucky-friday-111111.htm
5. This idea that people evaluate utility relative to a reference point was first proposed by Markowitz in 1952.
6. This paradox is attributed to Ellsberg (1961).
7. In their experiment, Tversky and Kahneman found that 88 percent of 200 subjects said yes to the first situation and 46 percent said yes to the second situation.
8. This is a very basic (and inadequate) description of discounting. The interested reader should examine hyperbolic discounting, which weights the future more heavily (see McCarty and Meirowitz 2007, 60).
9. For other alternative theories to expected utility theory, see Chew and MacCrimmon 1979; Machina 1987; and Quiggin 1982.

CHAPTER 5

SOLVING FOR A NASH EQUILIBRIUM IN NORMAL FORM GAMES

A. *IN COLD BLOOD*

Truman Capote's 1959 book *In Cold Blood* offers a true account of two convicts, Perry Smith and Dick Hickock, who were told in prison about a wealthy farmer in Kansas who kept a safe full of cash. Upon their release from prison, the two men drove to Kansas and robbed and murdered the owner, his wife, and their two 15- and 16-year-old children. The robbery netted the murderers less than 50 dollars. The police captured the two men in Las Vegas in about six weeks with evidence of their guilt for lesser crimes of parole violation and passing bad checks, but little physical evidence concerning the murders. As a result, the police placed Smith and Hickock in separate interrogation rooms to gain a confession. Prior to being arrested, each man had agreed to use a specified alibi if they were captured. However, once in the interrogation room, each of the suspects became uncertain about whether the other person would stick to his alibi. Smith thought that Hickock would be a convincing liar, but Hickock thought that Smith, being the more sensitive and guilt-ridden of the two, would lose his nerve and contradict the alibi. So the two murder suspects faced a dilemma. If both confessed to the crime and blamed it on the other, the district attorney would have enough evidence to sentence each to a life term, but not the death penalty. However, if one confessed and the other did not, the one who confessed would receive lesser charges and the other would get the death penalty. If both murder suspects stuck to their alibis, then they would only be convicted on the lesser crime for which police had hard evidence. So what do you think happened?

This story actually illustrates the classic game theory paradigm of the prisoner's dilemma, which I will model later in the chapter. The purpose of this chapter is to define the rules of game theory and start to analyze simple normal form games. In the next two sections, I will define game play and rules of the game, as well as formally define the concept of a Nash equilibrium. In the fourth section, I will solve the prisoner's dilemma game, and in the fifth section, I will explain the method of eliminating dominated strategies to find equilibrium in normal form games. In the sixth section, I will present a three-player normal form game followed by a spatial voting game in which elimination of dominated strategies leads to the median voter theorem. In the final section, I will present experiments that test this notion of dominated strategies.

B. BELIEFS AND THE COMMON KNOWLEDGE ASSUMPTION

A simple game can be defined by:

1. A set of players (p_1 to p_n)
2. A set of actions or strategies (a_1 to a_n)
3. Beliefs about the game environment (φ)
4. A choice of actions (a_1 to a_n) based on φ
5. A set of outcomes (x_1 to x_n)
6. A utility function that maps actions to an outcome $u_i(x)$

For example, consider a two-player game in which each player has two strategies and the selection of the joint strategies is converted by the utility function to yield an outcome x_i. Each player holds some beliefs about his or her payoffs in relation to his or her strategies, as well as beliefs about the other player's payoffs in relation to his or her strategies. The sequence of game play is as follows:

$$\left.\begin{array}{l}\text{Player 1 observes } (a_1, a_2) \text{ and } (a'_1, a'_2) \to \text{establishes } \varphi_1 \to \text{selects } (a_1, a_2) \to u_1(x) \\ \text{Player 2 observes } (a'_1, a'_2) \text{ and } (a_1, a_2) \to \text{establishes } \varphi_2 \to \text{selects } (a'_1, a'_2) \to u_2(x).\end{array}\right\} \to x_i$$

In this formulation, beliefs about complete information in simultaneous single-shot games are based on the common knowledge assumption, which means that players in a game know everything about the structure of the game, including all possible outcomes associated with each strategy. When common knowledge is assumed, then a player can understand the following regression: "I know it and you know it, but I know that you know it, and you know that I know it, and I know that you know that I know it, and you know that I know that you know it and..." Assuming common knowledge is akin to making all the parameters of the game public information.

In contrast, in incomplete information games we have to specify beliefs, since all parameters of the game are not known. For example, if a person confronts two unknown states and must make a choice in one state, then he has to have some beliefs about his current state, since he cannot make a choice in both states at the same time. This chapter and the next will be concerned with what occurs when beliefs are governed by the common knowledge assumption. I will define this notion of beliefs explicitly later in this book; for now, it is important to note that in game theory models, beliefs only relate to payoff-related information such as the probability of a person being in a particular state. In complete and perfect information games, the probability of being in a particular state is one.

C. NASH EQUILIBRIUM

1. DEFINING A NASH EQUILIBRIUM

Games can be represented in extensive form, which is a tree diagram representation, or in normal form, which is a matrix presentation. The extensive form is useful for modeling different information scenarios that a player might confront and will be examined later. Normal form games concern a situation in which players must make choices simultaneously and do not know each other's choices when making their own decisions.

Consider the generic normal form representation of a game displayed in Table 5.1. In this two-player game, Player 1 picks the row and selects Up or Down (the horizontal rows show that Player 1's payoffs for each strategy are on the left and go across), and Player 2 picks the column and selects Left or Right (the vertical columns show that Player 2's payoffs for each strategy are on the right and go down). The payoffs (4, 3, 2, 1, 0) are specified in each cell of the matrix. If Player 1 selects Up and Player 2 selects Left, then Player 1 would receive a payoff of 3 and Player 2 would receive a payoff of 2. In this game, Player 1 gets his highest payoff when he chooses Down and Player 2 selects Left, yielding a payoff of 4, and Player 2 gets her highest payoff when she chooses Right and Player 1 selects Up, yielding a payoff of 4. However, this is not the solution to the game.

Why can't both players achieve their highest payoffs simultaneously? Imagine that Player 1 selects Down to get his highest payoff of 4. We now have to determine whether Player 2 would be happy in this cell. If Player 2 is in the (Down, Left) cell, she will surely select her Right strategy to get a payoff of 1 (as opposed to 0). If both players are in the (Down, Right) cell, Player 1 gets a payoff of 2 and Player 2 gets a payoff of 1. Wouldn't Player 1 see that if he switched his strategy to the (Up, Right) cell, he would get a payoff of 3? But once in the (Up, Right) cell, Player 2 can only select Left, in which case she will receive a lower payoff of 2, and Player 1 can only move to his previous position, where he received a lower payoff of 2. Hence, neither player has an incentive to move out of the (Up, Right) cell, since by doing so he or she will receive a lower payoff. In this case, each player is selecting the best response to the other's strategy. An outcome of this nature is referred to as a Nash equilibrium.[1] A Nash equilibrium is an outcome in which, once inside a cell, neither player has a positive incentive to move out of the cell, since each player has selected his or her best response strategy to the other's strategies.

I would like to emphasize that a Nash equilibrium is not the highest payoff that both players can attain. Rather, the object of the game is for each player to achieve the highest possible payoff given the strategy the other player has selected. That is, a Nash equilibrium represents the best response strategy for one player against the best response strategy for another player. We will formally examine best response functions in Chapter 7, but for now, to convey the intuition behind this concept, I will introduce the notion of a Nash

TABLE 5.1   NASH EQUILIBRIUM

		Player 2	Column chooser
		Left	Right
Player 1	Up	3, 2	3, 4
Row chooser	Down	4, 0	2, 1

equilibrium more informally. In the previous game, Player 1 is unable to attain his highest payoff of 4 because of the strategic considerations of Player 2. Recall that a Nash equilibrium is the cell where each player selects his or her best strategy relative to an opponent's strategy. We can think of this cell as a kind of Monday morning quarterback's cell: if a player could watch a game unfold and observe the outcome and then had a chance to play the game again, would she play it differently? If the answer is no, then this is a Nash equilibrium.

2. NASH EQUILIBRIUM BEHAVIOR IN OTHER EXAMPLES

Consider the game in Table 5.2.

In this case, the Nash equilibrium appears in the (Down, Left) cell. Why is this a Nash equilibrium? If our players were in this cell and Player 1 had a chance to move, he would not, because he could only move up and receive a payoff of 4, which is less than 5. If Player 2 had an opportunity to move she also would not, since she could only move right and receive a payoff of -1, which is less than 0. Hence, no player has a positive incentive to move once in that cell. Now consider other cells and see whether there is an incentive for a player to select another strategy. If both players are in the (Up, Right) cell, Player 1 has an incentive to move down to get 6; if both players are in the (Down, Right) cell, Player 2 has an incentive to move left to get 0; and if both players are in the (Up, Left) cell, Player 1 has an incentive to move down to get 5. Consequently, we can define a Nash equilibrium as an outcome in which no player has a positive incentive to defect from his or her chosen strategy. When considering pure strategies, then, a game could have more than one Nash equilibrium or no pure strategy Nash equilibrium. However, when we discuss mixed strategies in Chapter 7, we will discover that all games of this type have an equilibrium in either pure or mixed strategies.

Notice that the definition of the Nash equilibrium requires players to have a positive incentive to move from a cell. What this means is that if a player is in a cell and can move

TABLE 5.2 ANOTHER NASH EQUILIBRIUM GAME

		Player 2	
		Left	Right
Player 1	Up	4, 4	4, 3
	Down	5, 0	6, −1

TABLE 5.3 EQUILIBRIUM WITH NO POSITIVE INCENTIVE TO MOVE AND MULTIPLE EQUILIBRIA

		Player 2	
		Left	Right
Player 1	Up	1, 1	0, 0
	Down	1, 2	0, 0

to a different cell that has the same payoff, then there is no positive incentive to move. Consider the following game in Table 5.3.

In this game, the (Up, Left) cell represents a Nash equilibrium, since Player 1 can only move down and will receive the same payoff of 1. Hence, Player 1 has no positive incentive to move. The (Down, Left) cell is also a Nash equilibrium, since Player 2 can only move right, which will yield her a lower payoff. In terms of behavioral game theory, this is an interesting game. Player 2's only choice is to move left. If the utilities in the game represent money, then Player 1 can move up, producing an even split for both players, or he can move down and give Player 2 twice the amount of money that he will receive. We can imagine that Player 1's selection would depend on his knowledge about his opponent. If he were playing against a friend, he might be willing to give his friend more money, but if he were playing against someone he did not like, then he would select the even-split option. But if he were playing against a stranger in an anonymous environment, how would he play? The outcome would depend on the psychological attributes of Player 1 in terms of his feelings about fairness, generosity, and so on. The outcome would also depend on the amount of money at stake. Player 1 might be more generous when the stakes were low but less generous when the stakes were high.

3. HE-THINKS-I-THINK REGRESS

As mentioned previously, not all games have a Nash equilibrium in pure strategies. Consider the game depicted in Table 5.4. How would each player play this game, since each wants to achieve his or her highest utility of 4? If the players are in the (Up, Left) cell, where Player 1 gets a utility of 4 and Player 2 gets a utility of 3, then Player 2 has an incentive to move right to the (Up, Right) cell, where she will get a utility of 4 and Player 1 will get a utility of 1. Knowing this, Player 1 has an incentive to change his choice from up to down and move to the (Down, Right) cell, increasing his utility to 3. Player 2 is then confronted with a utility of 1, so she has an incentive to switch her choice from right to left and move to the (Down, Left) cell to get a utility of 2. This progression is called the "He-thinks-I-think regress."[2] Each player has an incentive to change his or her choice when he or she knows what the other player is going to choose. But since the game is played simultaneously and with common knowledge, then neither player knows how the other player will play before selecting his or her strategy. Hence, there is no equilibrium in pure strategies for this game. In Chapter 7, I will discuss how to calculate an equilibrium for this game using mixed strategies.

TABLE 5.4   HE-THINKS-I-THINK REGRESS

		Player 2	
		Left	Right
Player 1	Up	4, 3	1, 4
	Down	2, 2	3, 1

4. PARETO PRINCIPLE

One of the primary exercises of game theory is finding solutions or equilibria to games. Considering the previous game, you might think that the (4, 3) cell would be a reasonable solution, since this combined cell offers the best outcome for both of the players. However, our definition of a Nash equilibrium prevents us from deeming this cell the outcome, since if Player 2 thinks that Player 1 is reasonable and that Player 1 finds that cell to be reasonable, then Player 2 has no other choice but to "selfishly" move right and get 4 for herself. "Pareto optimality" is often referenced as a way to model this situation. An outcome is considered to be Pareto suboptimal if there is another outcome that makes both (or all) players better off. A Pareto optimal outcome is one in which no further improvement is possible and any shift from a cell decreases the payoff of at least one player. In the game in Table 5.4, the (Up, Left) outcome is the Pareto optimal solution, since it offers the best outcome that both players can achieve, even though Player 1 receives a better payoff (4) than Player 2 (3). Game theory seeks to understand ways in which players can deviate from a Nash equilibrium and achieve an outcome that is Pareto optimal (i.e., an outcome that is better than the Nash equilibrium for all players).

5. NASH EQUILIBRIUM IN A ZERO-SUM GAME

Thus far we have only considered non-zero-sum games, but zero-sum games are no different and can also be solved for a Nash equilibrium. Recall that zero-sum games are games in which the sum of payoffs across players is zero. Be aware that since zero-sum games involve negative numbers, the direction of relationships are inversed from what we normally consider and can create counterintuitive confusion.

Consider the zero-sum game in Table 5.5. In this game, a unique equilibrium can be found in the (Down, Left) cell. Once in this cell, Player 1 has no incentive to defect, since by defecting he can only get –1, and Player 2 has no incentive to defect, since she can only decrease her payoff to –4. An examination of other cells can confirm that the (Down, Left) cell is the Nash equilibrium for this game.

D. PRISONER'S DILEMMA

Recall the prisoner's dilemma example in the introduction in which the two suspected murderers have been arrested and the district attorney separates them and offers each the same deal. The deal is that if one criminal reveals the details of the crime before the other, then the

TABLE 5.5 ZERO-SUM GAME

		Player 2	
		Left	Right
Player 1	Up	–1, 1	–3, 3
	Down	2, –2	4, –4

one who confesses first will receive a lesser sentence and the suspect who does not confess will get the death penalty. If both suspects confess, they will each get life in prison. However, if neither confesses, they will only be convicted on the lesser charges. The latter deal is the best (or Pareto optimal) outcome, and both criminal suspects should remain silent and not confess. However, because of the construction of payoffs in this game, the equilibrium is for both players to confess. The dilemma in this game is that both players have an incentive to confess, producing an outcome that is worse than the outcome of both remaining silent. This game illustrates the problem of cooperation: both players would be better off if they could be "forced" to be cooperative, but because there is no way to enforce a binding agreement, both have an incentive to not cooperate and "rat" on each other. The normal form of this game is presented in Table 5.6.

Equilibrium occurs when both players have a dominant strategy to confess. A dominant strategy is a strategy for which the value of payoffs is higher than (or equal to) the value of the payoffs of an alternative strategy. That is, it is the best strategy a player can play regardless of what the opponent plays. If the opponent has a dominant strategy, then she is forced to play that strategy, since it is the best strategy she has regardless of what her opponent plays. Since both players are playing their best or dominant strategy, then these represent their best responses against each other and produce a Nash equilibrium. In the prisoner's dilemma game depicted in Table 5.6, notice that the values of Player 1's strategy to confess (given both strategies for Player 2) are –5 and –20, which are both higher than the values of his no-confess strategy (i.e., –5 > –10 and –20 > –30). Hence, the confess strategy dominates the no-confess strategy. The same is true for Player 2: the values of her confess strategy are –5 and –20, which are higher than the values of her no-confess strategy (–10 and –30). As a result of the payoff configuration, and because rationality dictates that players must select their highest possible payoff, both players are duped into selecting their dominant strategies and end up in the equilibrium, in which both confess.

The "dilemma" in the prisoner's dilemma is: How can both players cooperate and end up in the no-confess cell, where they are both better off? The structure of the game makes this solution impossible unless some enforcement mechanism or incentive scheme is inserted into the game, since both players have dominant strategies to confess. This problem is also characteristic of social dilemma games, such as a public goods game that aims to get people to cooperate. For example, if we substitute "contribute" for "no-confess" and "refuse to contribute" for "confess," then we can see that players have no incentive to contribute, since free-riding is the dominant strategy. I will discuss social dilemma games in more detail in the next chapter.

TABLE 5.6   PRISONER'S DILEMMA

		Player 2	
		No confess	Confess
Player 1	No confess	−10, −10	−30, −5
	Confess	−5, −30	−20, −20

E. ELIMINATION OF DOMINATED STRATEGIES AND A DOMINANT SOLVABLE EQUILIBRIUM

Before we begin this section I would first like to define some terms related to how one strategy can "beat" another strategy. By beat, I simply mean that one strategy is better than another strategy. If a player has a list of strategies, then he will select the best one to play. One way that he can do this is to eliminate all the strategies he knows he will not play because some other strategy is better. A dominant strategy beats another strategy, a dominated strategy is beaten by another strategy, and a dominate strategy beats all other strategies. When games have dominant strategies, it is possible to search for and eliminate those strategies that are dominated. If one strategy beats all other strategies, then we call it a dominate strategy.

This process of searching for and eliminating strategies that are dominated is a way in which players arrive at their best response to the strategy that another player might play. Think of a player who has two strategies, Up and Down, with each strategy having three elements (or payoffs). Table 5.7 depicts Player 1's payoffs.

Notice that all the payoffs for the Down strategy are higher than the payoffs for the Up strategy. That is, 4 is greater than 1, 6 is greater than 2, and 6 is greater than 3. Therefore, Player 1 will always do better by selecting the Down strategy.

Strategies can also be weakly dominant if one or more elements of two or more strategies have the same payoff. Consider Table 5.8. In this case, the Down strategy weakly dominates the Up strategy, since it is at least as good as the Up strategy and, depending on what Player 2 does, could be better.

The process of eliminating dominated strategies consists of removing all strategies that are dominated or weakly dominated by another strategy. If an iterated elimination procedure ends in only one strategy per player, the game is considered dominant solvable. Consider the game depicted in Table 5.9. In this game Player 1's Down strategy dominates his Middle strategy, since elements of Down (4, 3, 5) are greater than elements of Middle (3, 2, 4). Hence, we can eliminate Player 1's Middle strategy. See Table 5.10 for the resulting table.

TABLE 5.7   PLAYER 1'S DOMINANT STRATEGY

Player 1	Up	1	2	3
	Down	4	6	6

TABLE 5.8   PLAYER 1'S WEAKLY DOMINANT STRATEGY

Player 1	Up	1	2	3
	Down	1	2	4

TABLE 5.9   DOMINANT ELIMINATION GAME

		Player 2		
		Left	Middle	Right
	Up	5, 6	5, 5	6, 3
Player 1	Middle	3, 2	2, 3	4, 2
	Down	4, 3	3, 3	5, 3

In fact, Player 1's Down strategy is dominated by his Up strategy, since all elements of Up (5, 5, 6) are greater than the elements of Down (4, 3, 5). We can thus eliminate the Down strategy. Table 5.11 shows the results of this second elimination.

Since Player 1 will certainly play an Up strategy, then Player 2 has no incentive to play a Middle or Right strategy, since Left yields the greatest payoff. Consequently, this game is dominant solvable with the (Up, Left) cell being the Nash equilibrium. See Table 5.12 for this result.

It is important to understand that when players are checking for and removing dominated strategies, they are checking for their best responses to what another player can play. Rasmussen (1989, 33) notes:

> Every dominant-strategy equilibrium is a Nash equilibrium, but not every Nash equilibrium is a dominant-strategy equilibrium. If a strategy is dominant it is a best response to *any* strategy the other players pick, including their equilibrium strategies. If a strategy is part of a Nash equilibrium, it need only be a best response to the other players' *equilibrium* strategies.

Let us consider another game, depicted in Table 5.13, which is not dominant solvable.

TABLE 5.10   FIRST ELIMINATION

		Player 2		
		Left	Middle	Right
	Up	5, 6	5, 5	6, 3
Player 1	~~Middle~~	~~3, 2~~	~~2, 3~~	~~4, 2~~
	Down	4, 3	3, 3	5, 3

TABLE 5.11   SECOND ELIMINATION

		Player 2		
		Left	Middle	Right
	Up	5, 6	5, 5	6, 3
Player 1	~~Middle~~	~~3, 2~~	~~2, 3~~	~~4, 2~~
	~~Down~~	~~4, 3~~	~~3, 3~~	~~5, 3~~

TABLE 5.12   DOMINANT SOLVABLE EQUILIBRIUM

		Player 2		
		Left	Middle	Right
	Up	5, 6	~~5, 5~~	~~6, 3~~
Player 1	Middle	~~3, 2~~	~~2, 3~~	~~4, 2~~
	Down	~~4, 3~~	~~3, 3~~	~~5, 3~~

TABLE 5.13   SECOND DOMINANT ELIMINATION GAME

		Player 2		
		Left	Middle	Right
	Up	0, 1	−2, 3	4, −1
Player 1	Middle	0, 3	3, 1	6, 4
	Down	1, 5	4, 2	5, 2

First, Player 1 eliminates the Up strategy, because it is dominated by the Down strategy, since the elements of Down (1, 4, 5) are all greater than the elements of Up (0, –2, 4). Table 5.14 displays the results of this elimination.

Next, we can eliminate the Middle strategy for Player 2, which is dominated by both the Left and Right strategies. The subsequent results are displayed in Table 5.15.

Consequently, we are left with the following 2 × 2 game, shown in Table 5.16, which is not dominant solvable. This game has two Nash equilibria: (Down, Left) and (Middle, Right).

F. THREE-PLAYER NORMAL FORM GAMES

Thus far we have only been concerned with two-player games; however, normal form games can also be used to model three or more players. I will not examine games involving more than three players because they can be messy to present in tabular form. Consider the three-person game presented in Table 5.17. As the arrows indicate, Player 1 controls the rows and can move either up or down, Player 2 controls the columns and can move right or left, and

TABLE 5.14 FIRST ELIMINATON

		Player 2		
		Left	Middle	Right
Player 1	Up	~~0, 1~~	~~–2, 3~~	~~4, –1~~
	Middle	0, 3	3, 1	6, 4
	Down	1, 5	4, 2	5, 2

TABLE 5.15 SECOND ELIMINATON

		Player 2		
		Left	Middle	Right
Player 1	Middle	0, 3	~~3, 1~~	6, 4
	Down	1, 5	~~4, 2~~	5, 2

TABLE 5.16 REMAINDER GAME WITH TWO EQUILIBRIA

		Player 2	
		Left	Right
Player 1	Middle	0, 3	6, 4
	Down	1, 5	5, 2

TABLE 5.17 THREE-PLAYER GAME

		Player 3			
		Right Table		Left Table	
		Player 2			
		Right	Left	Right	Left
Player 1	Up	15, 15, 15	11, 21, 11	10, 9, 20	10, 10, 10
	Down	22, 11, 12	12, 12, –1	13, 9, 0	15, 10, 0

Player 3 controls the tables and can select the left or right table. Within each cell, the first payoff is associated with Player 1, the second payoff is associated with Player 2, and the third payoff is associated with Player 3.

Three-player games can also be solved by the process of eliminating dominated strategies. First consider Player 1, for whom all elements of the Down strategy (22, 12, 13, 15) are greater than all elements of the Up strategy (15, 11, 10, 10). We can thus eliminate this strategy. Results are displayed in Table 5.18.

Player 2 must now select between the Right and Left strategies. Because both elements of his Right strategy (11, 9) are dominated by the elements of his Left strategy (12, 10), we can eliminate Player 2's Right strategy.

Now Player 3 has to decide between the Right table and the Left table. Since 0 is greater than –1, she will select the Left table, making (Down, Left, Left) the equilibrium to this game.

This type of informal strategy analysis will allow us to think about the meaning of the best response strategy that will be discussed in Chapter 7. The idea of a best response strategy is to select the best available strategy relative to an opponent's available strategies. A dominant strategy is by definition a best response to an opponent's strategy, since it is the best that a player can do in response to whatever strategy the opponent plays.

G. ELIMINATING DOMINATED STRATEGIES IN AN ELECTION GAME

To illustrate the concept of dominant strategies in a different context, I will examine a simple election game. Consider two candidates who are competing for some political office by selecting policy positions over a single dimension (the x-axis) that ranges from 0 to 100, as in Figure 5.1. Assume that there are three voters who have single-peaked utility functions so

TABLE 5.18 FIRST ELIMINATION OF THREE-PERSON GAME

		Player 3			
		Right		Left	
		Player 2		Player 2	
		Right	Left	Right	Left
Player 1	Up	~~15, 15, 15~~	~~11, 21, 11~~	~~10, 9, 20~~	~~10, 10, 10~~
	Down	22, 11, 12	12, 12, –1	13, 9, 0	15, 10, 0

TABLE 5.19 SECOND ELIMINATION OF THREE-PERSON GAME

		Player 3			
		Right		Left	
		Player 2		Player 2	
		Right	Left	Right	Left
Player 1	Up	~~15, 15, 15~~	~~11, 21, 11~~	~~10, 9, 20~~	~~10, 10, 10~~
	Down	~~22~~, 11, 12	12, 12, –1	~~13~~, 9, 0	15, 10, 0

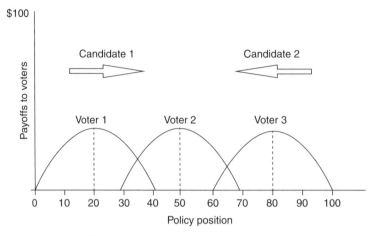

FIGURE 5.1   Median voter theorem

that they vote for the candidate who is closest to them on the dimension. Notice that the y-axis measures the payoffs to voters in terms of the height of their utility function. If Candidate 1 selects position 20 and Candidate 2 selects position 60, then Voter 1 will vote for Candidate 1, since that candidate's position is closer to Voter 1's ideal point than is the other candidate's position. Voters 2 and 3 will vote for Candidate 2, since 60 is closer to their ideal points, and this candidate will win.

At position 20, Candidate 1's strategy is dominated by Candidate 2's. Knowing this, Candidate 1 moves her position to 41. Now Candidate 2's position of 60 is dominated by Candidate 1's position of 41, since Voters 1 and 2 prefer that position to the position of 60. Candidate 2 sees that Candidate 1 has switched positions, so he moves away from his dominated strategy to 50. In response, Candidate 2 moves away from her dominated strategy to 50. By eliminating all dominated strategies in this game, both candidates adopt 50 as their "weakly" dominant strategies (since they tie) and we establish the famous median voter theorem (Downs 1957; Black 1958).[3] This theory claims that in a two-candidate election that takes place over a single dimension, when there is an odd number of voters who have single-peaked preferences and are all fully informed, then candidate positions will converge to the median voter's ideal point. This example illustrates the tendency in a two-party system for parties to adopt ideological policies that are either moderate right or moderate left (near the median voter's ideal point).

H. FINDING DOMINATE STRATEGIES IN A SPATIAL ELECTION EXPERIMENT

An experiment by McKelvey and Ordeshook (1985) indirectly tests this notion of eliminating dominated strategies in a spatial voting context. In this experiment, candidates were given incomplete information about the voters' locations and voters were given incomplete information about the candidates' positions.[4] We will discuss the notion of incomplete information in considerable detail later, but for now, just think of this as an experiment in which all subjects have some location on a single dimension and know their own location, but not the location of other subjects, on the same dimension. One question would then be: Within this uncertain environment, can subjects find and locate strategies that are dominated by other strategies and eliminate them to find the dominate strategy?

In this experiment, subjects participated in the election environment described previously, in which two candidates adopted positions on a single dimension and each voter subject was assigned a single-peaked utility function that determined their preferences for candidates (see Figure 3.9). Voter subjects were divided between informed voters who knew the positions that candidates adopted and uninformed voters who did not know the positions. Uninformed voters were only told which candidate position was farthest to the left of their position. Candidate subjects did not know where voters were located on the dimension. The median point for uninformed voters was at 45–48, the median for informed voters was at 75, and the overall median was at 60. Recall that the experiment aimed to determine whether candidate subjects could locate the overall median.

Prior to the election, subjects participated in two polls. McKelvey and Ordeshook (1985) developed a model in which informed voters would vote correctly in the first poll and uninformed voters would vote randomly. In the second poll, uninformed voters should be able to decipher the candidate positions from candidates' left-right orientation and the vote distribution of the first poll. McKelvey and Ordeshook thus argued that uninformed voters should behave as informed voters and candidates should converge to the overall median. To arrive at these outcomes, candidate subjects had to locate and eliminate dominated strategies to find the location that was desirable to the most voters. Subjects participated in 10 rounds of the experiment.

The experiment showed that about two-thirds of the uninformed voters behaved as if they were fully informed. After the fourth round of the experiment, candidate positions converged near the overall median of 60. These experiments provided evidence that the full information assumption can be relaxed in this kind of simple model, because candidate subjects can still successfully eliminate dominated strategies to find the dominant strategy, or the median voter's ideal point. While this experiment shows that dominate strategies can be found and played within this context, in other contexts this is not the case, as the next experiments illustrate.

I. OTHER EXPERIMENTAL TESTS OF DOMINANT STRATEGIES

1. TVERSKY AND KAHNEMAN'S DOMINANT STRATEGY EXPERIMENT

Game theory assumes that players can identify and successfully eliminate strategies that are dominated. Tversky and Kahneman (1986b, 263–64) conducted an experimental test of whether subjects can actually identify dominant strategies.[5] In this experiment subjects were given two hypothetical problems as follows:

> Consider the following two lotteries, described by the percentage of different colored marbles in each of two boxes and the amount of money you win or lose depending on the color of a randomly drawn marble. Which lottery do you prefer?

TABLE 5.20   TWO LOTTERIES

Option A	90% white, $0	6% red, $45	1% green, $30	1% blue, –$15	2% yellow, –$15
Option B	90% white, $0	6% red, $45	1% green, $45	1% blue, –$10	2% yellow, –$15

TABLE 5.21   ALTERNATE VERSION OF TWO LOTTERIES

Option C	90% white, $0	6% red, $45	1% green, $30	3% yellow, –$15
Option D	90% white, $0	7% red, $45	1% green, –$10	2% yellow, –$15

In this lottery, displayed in Table 5.20, Option B clearly dominates Option A, and all the subjects in the experiment consequently chose Option B. In another treatment, Tversky and Kahneman attempted to hide the dominant strategy by combining the red and blue choices in Option B and the yellow and blue choices in Option A. In this new lottery, displayed in Table 5.21, subjects were again asked which lottery they would prefer.

Although the lotteries were the same, subjects selected Option C over Option D. Why? As in the previous lotteries, the expected value for D is greater than C, but subjects appeared to respond to the number of wins in the lottery, for which C had two positive outcomes and one negative outcome and D had two negative outcomes and one positive outcome. Tversky and Kahneman concluded that subjects will detect dominant strategies when they are transparent, but these strategies can easily become unrecognizable depending on the context (or frame) of the game.[6] The economists Davis and Holt (1993) note that subjects in second-priced auctions often fail to identify dominant strategies because they are not transparent. But Davis and Holt suspect that if subjects know what to look for (i.e., they are schooled on this type of auction), they will be able to identify these strategies easily.

2. BEAUTY CONTEST

The beauty contest or guessing game is another way of showing the inability of subjects to select weakly dominant strategies. This game, devised by John Maynard Keynes (1936), illustrates the iterated reasoning needed to identify dominant strategies by imagining a contest in which people select among the faces of six beauty queens. Keynes notes (1936, 156):

> It is not a case of choosing those which, to the best of one's judgment, are really the prettiest, nor even those which average opinion genuinely thinks the prettiest. We have reached the third degree where we devote our intelligences to anticipating what average opinion expects the average opinion to be. And there are some, I believe, who practice the fourth, fifth and higher degrees.

In this game, two people simultaneously choose a number between 0 and 100, and the winner is the person who selects the number closest to two-thirds of the average of the two numbers.[7] For example, if one person selects 10 and another person selects 20, the average of those two numbers is 15, and two-thirds of 15 is 10. Therefore, the person who selected 10 would win. The weakly dominant strategy for this game is zero, since the person who selects the lowest number wins. However, when subjects play this game in experiments, the weakly dominant strategy is rarely selected. To understand why, Camerer (2003, 17) explains:

> Most players start by thinking, "Suppose the average is 50." Then you should choose 35, to be closest to the target of 70 percent of the average and win. But if you think all players will think this way the average will be 35, so a shrewd player such as yourself (thinking one step ahead) should chose 70 percent of 35, around 25. But if you think all players think that way you should choose 70 percent of 25, or 18.
>
> In analytical game theory, players do not stop this iterated reasoning until they reach a best-response point. But, since all players want to choose 70 percent of the average, if they all choose the same number it must be zero.

One explanation for this result is cognitive bias, which leads subjects to underestimate the value of their own choice in a game (Nagel 1998). Other experiments argue that subjects simply do not understand the game form. Chou et al. (2007) conducted experiments that show that if subjects are given a hint in the game (e.g., they are told that the lowest number wins), then they have a better understanding of the game and select weakly dominant strategies more often.

These results show that when people have to search for strategies, they often have a difficult time finding the optimal one to use. However, different environments appear to make the search for these strategies easier. This issue is important, because people in real-life situations must often search for strategies when making a decision. Most game theory models assume that strategies simply exist and are not searched for. The learning models discussed in the last chapter are relevant to study this behavior.

What these experiments show is that dominant strategies might exist in a game, but players do not always know they are there. It is unclear whether framing, psychological factors, or understanding of the game form contribute to subjects' inability to play dominant strategies when they are present in a game. Camerer (2003, 17) comments:

> In practice, it is unlikely that people perform more than a couple of steps of iterated thinking because it strains the limits of working memory (i.e., the amount of information people can keep active in their mind at one time). Consider embedded sentences such as "Kevin's dog bit David's mailman whose sister's boyfriend gave the dog to him." Who's the "him" referred to at the end of the sentence? By the time you get to the end, many people have forgotten who owned the dog because working memory has only so much space. Embedded sentences are difficult to understand. Dominance-solvable games are similar in mental complexity.

J. SUMMARY

In this chapter I laid out the structure of a basic game and demonstrated how to calculate and search for a Nash equilibrium. One way to search for a Nash equilibrium is to determine whether a player has a dominant strategy by eliminating all dominated strategies. This process allows players to search for their best response to each other. I also examined an election game in which candidates remove dominated strategies to discover the weakly dominant strategy for each candidate. This behavior results in equilibrium that is associated with the median voter's ideal point. Experiments have shown that subjects are able to identify dominated strategies when they are obvious, but have a difficult time when they are hidden or involve complex calculations.

NOTES

1. This equilibrium concept is named after John Nash, a professor at Princeton University who won the Nobel Prize for his contribution to game theory and was the subject of the book (and later movie) entitled *A Beautiful Mind*.
2. A number of movie scenes illustrate this regression, with one of the best being the poisoned drink scene in *The Princess Bride*.
3. This theory was first introduced by Harold Hotelling in 1929.
4. This experiment is important in establishing that candidates can find equilibrium positions and voters can identify candidate positions when all have incomplete information. These results loosened the information restrictions of the standard spatial voting model of elections. See Morton and Williams (2011) for a review of this literature.
5. It should be noted that Tversky and Kahneman's definition of dominance is different from the definition specified here. My definition requires that there be two or more players, but in these experiments there is no opponent.
6. According to the authors, this example also illustrates a violation of the assumption of reduction of compound lotteries.
7. This is often called a beauty contest game because its original formulation revolved around ranking the beauty of six women's faces (see Keynes 1936).

CHAPTER 6

CLASSIC NORMAL FORM GAMES AND EXPERIMENTS

A. CLASSIC NORMAL FORM GAMES

In this chapter I will discuss four classic normal form games: the prisoner's dilemma, the chicken game, the battle of the sexes, and the stag hunt. These games represent broader areas of study such as repeated interactions, social dilemma games, brinkmanship games, coordination games, and assurance games. For each area I will present experiments that illustrate how these games have been empirically tested. I would like to stress that this is not a literature review of all the experiments or theoretical findings in a particular area. That type of analysis is beyond the scope of this introductory book. For example, the prisoners' dilemma game is by far the most studied game today in both theoretical and experimental research. A mere review of all the variants of prisoners' dilemma experiments would itself fill a two-volume text. The goal of this chapter is to show how these seemingly specific games address a more general problem in the social sciences and illustrate how these games have been experimentally tested.

Before I proceed I would like to recall our definition of an experiment from Chapter 2. Many of the experiments in this chapter, which are taken from varied academic disciplines, would not fit a narrow definition of an experiment, since they do not involve randomizing subjects into different treatments. However, they do fulfill the requirement of our broader definition that an experiment manipulate the DGP (data-generating process).

In the next section I will revisit the prisoner's dilemma game to discuss its repeated nature and social dilemmas. I will then examine the chicken game and the dove-hawk game, which present an ideal situation of brinkmanship. In the fifth section I will present the battle of the sexes game and highlight the problem of coordination. In the last section I will illustrate the stag hunt, which models assurance games.

B. REVISITING THE PRISONER'S DILEMMA

1. REPEATED PRISONER'S DILEMMA GAME

A stage game is just another name for a normal form game. In a repeated game, players play the same stage game a number of times, with its actions and values typically remaining fixed from period to period.[1] The first experiment of the prisoner's dilemma game was a stage game that was played 100 times by two researchers, John D. Williams of the RAND Corporation and Armen Alchian of the University of California, Los Angeles, in the early 1950s (Poundstone 1992, 106). Recall that the dilemma of this game is that players have incentives to not cooperate even though cooperation is in both players' best interest. But what happens if this game is played many times with the same players? In this case, a player's reputation begins to factor into how players play the game; that is, players reveal a pattern of strategy choice to each others, developing their reputations. A player's reputation affects reciprocity rates in the sense that cooperative play by one player may increase cooperative play by another and noncooperative play may induce other noncooperative play. In a repeated game, a player often selects a strategy in the present in reaction to a series of past strategies selected by another player.

Consider the stage game displayed in Table 6.1, in which the equilibrium outcome is for both players to defect and receive a payoff of (3, 3), as opposed to a cooperative payoff of (5, 5). But what happens if this game is played for a finite number of rounds and each player knows how many rounds will be played? As we see in Chapter 8, one way to solve this game is by using backward induction, a technique that lets us see what happens in the last round of the game and work back to the start of the game. Since this is a finite game of complete and perfect information, we can figure out each player's rational strategy in each round and determine the outcome. By starting at the last move of the game, we can figure out the optimal action to take and then use this information to infer what action should be taken on the penultimate move, and so on to the first move. In a finite repeated prisoner's dilemma, there is no benefit to cooperating on the last round of the game, so both players have an incentive to defect. Since there is no incentive to cooperate on the last move, then players know that cooperation will not be achieved, and so they should also not cooperate on the penultimate move. This logic can be extended so that the equilibrium is for players to defect every round. However, this calculation assumes that each play of the stage game is an independent event that does not depend on reputations, but experimental and observational data tell us that reputational effects can sustain cooperation. Hence, one question that game theory seeks to answer is what types of mechanisms enhance cooperation in settings where cooperation is difficult to achieve. For example, in a repeated game in which players are expected to interact with others over some period of time, players' behavior deviates from the Nash equilibrium. Why do they deviate from rationality? These deviations occur for a myriad of psychological reasons related to social preferences, as we discussed in Chapter 3.

TABLE 6.1   PRISONER'S DILEMMA AS A REPEATED GAME

		Player 2	
		Cooperate	Defect
Player 1	Cooperate	5, 5	0, 10
	Defect	10, 0	3, 3

2. EXAMPLE OF FINITE REPEATED GAME WITH RECIPROCITY STRATEGIES

In finite repeated games, players play strategies that are a function of the strategies that another player is playing (or is anticipated to play) and also take into account the history of strategies that an opponent has played over some time period.[2] These strategies are called reciprocity strategies. Observing an opponent's strategy history allows a player to use more informed strategies. There are a multitude of strategies that can be played in a repeated game, but the following list offers a few of the most popular. Strategies c and d are reciprocity strategies, in that they are a function of the strategy chosen by an opponent.[3]

a. *All C's*: Both players cooperate on every round and their payoffs at the end of the game are 50, 50.
b. *All D's*: Both players defect on every round and their payoffs at the end of the game are 30, 30.
c. *Tit for tat*: Players cooperate and match each other's moves. That is, if one player cooperates on a round, the other player cooperates, and if one defects, the other also defects.
d. *Grim strategy*: Players cooperate on the first round and once one player uses a defection strategy, the other player defects every round thereafter.

In a repeated game a player may use a combination of these strategies throughout game play. To illustrate how repeated games can progress, consider the following 10-period game in Table 6.2, which corresponds to the game outlined in Table 6.1 in which players know they are playing only 10 rounds. To determine the final payoff for each player, we simply add up the payoffs for each round (e.g., if a player defects each period his total payoffs would be 30, and if he cooperates each period his total payoffs would be 50).

Who wins? It is a tie, with payoffs of 48/48. Consequently, each player would be better off cooperating (to receive a payoff of 50) rather than engaging in warfare, as in Rounds 6 through 10.[4] In the Alchian/Williams experiment, one player selected cooperation 68 percent

TABLE 6.2 EXAMPLE OF A HYPOTHETICAL FINITE PRISONER'S DILEMMA GAME

Round	Action (P1, P2)	Payoffs (P1, P2)	Comments
Round 1	C, D	0, 10	P2 wants to win so P2 defects
Round 2	D, C	10, 0	P1 punishes P2 for defection
Round 3	C, C	5, 5	P1 and P2 cooperate
Round 4	C, C	5, 5	P1 and P2 are coexisting
Round 5	C, C	5, 5	P1 and P2 are happy
Round 6	D, C	10, 0	P1 tires of cooperating so she defects
Round 7	C, D	0, 10	P1 wants peace but P2 retaliates
Round 8	D, C	10, 0	P1 retaliates backs
Round 9	C, D	0, 10	P2 retaliates back
Round 10	D, D	3, 3	Last round so, both defect

of the time and the other player selected defection 78 percent of the time. Mutual defection (the Nash equilibrium) only occurred 14 times (Poundstone 1992). Poundstone (1992, 116) notes:

> When [the results were shown] to Nash, he objected that "the flaw in the experiment as a test of equilibrium point theory is that the experiment really amounts to having players play one large multi-move game. One cannot just as well think of the thing as a sequence of independent games as one can in zero-sum cases. There is too much inter-action, which is obvious in the results of the experiment."
>
> This is true enough. However, if you work it out, you find that the Nash equilib-rium strategy of the multi-move "supergame" is for both players to defect in each of the hundred trials. They didn't do that.

Again, the importance of this example is that it shows that, contrary to a single-shot model for which defection is the dominate strategy, cooperation can be sustained in repeated games.

3. AXELROD'S TOURNAMENT

Axelrod (1984) conducted a computer tournament in which he invited 14 scholars from diverse academic backgrounds (including game theorists) to play a repeated prisoner's dilemma game against each other.[5] In this tournament, participants wrote down a set of rules (strategies) that were matched against other participants' strategies. The tournament was conducted in a round-robin fashion, so that each entry was paired with every other entry. Each game consisted of 200 rounds. The strategy of "tit for tat" won the tournament, but other strategies that relied on being nice (or never defecting first) also did well. These results suggest that emotions such as "niceness" help induce cooperation. Discussing the failed strategies, Axelrod (1984, 40) comments:

> Even expert strategists from political science, sociology, economics, psychology, and mathe-matics made the systematic error of being too competitive for their own good, not being for-giving enough, and being too pessimistic about the responsiveness of the other side.... The effectiveness of a particular strategy depends not only on its own characteristics, but also on the nature of the other strategies.

The repeated prisoner's game is a good game to use when studying emotions since, as Axelrod notes, the strategies it uses invoke emotional behavior and responses.

In more controlled experiments testing 2×2 finite, repeated prisoner's dilemma games in which players possess full information about their opponent's strategies and payoffs, high levels of cooperation have also been shown to be sustained for prolonged periods (Rapoport and Chammah 1965). Other experiments that have relaxed these conditions have also found lower levels of cooperation (see Van Huyck, Wildenthal, and Battalio 1992). There-fore, one way to partially solve the dilemma and induce cooperation is to have repeated play of the stage game. Social interaction causes people to cooperate, even if it is not in their self-interest to do so.

4. PRISONER'S DILEMMA AS A ROUTE-CHOICE GAME

Before I abandon repeated games I want to present an interesting application called the route-choice game, which examines traffic congestion (Helbing et al. 2005). In this game, players have a choice between Route 1 and Route 2, where Route 1 is a highway and Route 2 is a side street that goes to the destination but is slower than Route 1. The dilemma in this game is that if everyone selects Route 1, there will be traffic congestion and the trip will be slower than taking Route 2.[6]

In a game with 10 players, it is best for a single player to take Route 1 and for the others to take Route 2, so that there will be no traffic congestion and the player will get to her destination with ease. The worst strategy for all players is for all of them to select Route 1, which will result in a traffic backup that slows down everyone's trip. The optimal solution depends on the capacity of the side streets, so the 10 players should cooperate, with half taking Route 1 and half taking Route 2. In a two-person experiment that examined an iterated version of this game (Helbing et al. 2005), subjects played 300 rounds, for which the optimal solution was for them to cooperate and alternate routes each round.[7] Hence, in one round, Player 1 selected Route 1, in the next period she selected Route 2, and so on. The optimal solution was generally found, but it took subjects 220 rounds of game play to discover it.

C. SOCIAL DILEMMAS

1. COLLECTIVE GOODS PROBLEM

Prisoner's dilemma games have also been used to model social dilemmas such as the collective goods problem (Hardin 1982; Ostrom 2000). A collective good has two properties: First, the benefits from the good are nonexcludable—once produced, everybody can enjoy the benefits for free. Second, the supply of the good is inexhaustible (jointness of supply), so that if an individual uses it, it is still fully available to everyone else. For example, public radio is funded in part by listeners' contributions. However, you do not have to contribute to public radio to listen to the programs. All you have to do is tune a radio to the correct frequency (nonexcludable nature). As well, if one person listens to the program, this act does not diminish others' capacity to listen (jointness of supply). People who do not pay for the public good but enjoy its benefits are referred to as free-riders. To see how a prisoner's dilemma game can represent the collective action problem, consider the game illustrated in Table 6.3.

Let us assume that the payoffs represent levels of some public good and the payoff of 4 represents the status quo. If neither of two players contributes to the public good, the status

TABLE 6.3   PRISONER'S DILEMMA AS COLLECTIVE GOODS PROBLEM

		Player 2	
		Contribute	Don't contribute
Player 1	Contribute	5, 5	3, 6
	Don't contribute	6, 3	4, 4

quo prevails. If both players decide to contribute, they attain a level of public good that is greater than the status quo. If one player decides to contribute and the other does not, then the player who does not contribute receives the contribution to the public good provided by the other player and does not have to pay the cost, so he free rides and attains his highest utility. Although the player who contributes gets the lowest payoff (since she pays the cost), without the funds from the other player, the public good is diminished below the status quo point.

2. COLLECTIVE GOODS EXPERIMENT

Laboratory experiments have empirically tested this collective goods game (see Andreoni and Miller 1993; Issac, Walker, and Williams 1994; and Ledyard 1995). In a typical public goods experiment, five subjects are placed in a room and given an endowment in the form of an initial sum of cash, such as 10 one-dollar bills. Subjects are told that they can use their endowments to invest in a group fund or a common pool. They are allowed to invest any dollar amount between zero and 10 in the group fund. Subjects are then given an envelope in which they can place as many dollars as they choose. The experimenter collects all the envelopes and places the money they contain in the group fund; he then doubles the amount within the group fund and splits it equally among all the subjects. Only the experimenter knows how much each subject has contributed, and the subjects are unaware of how much others have contributed. All they know is the amount of the total contributions received.

The prediction is that no subject should contribute to the group fund and they should all go home with 10 dollars. However, if everyone contributes 10 dollars to the fund, then each person can go home with 20 dollars. But say that one subject contributes 10 dollars and no one else contributes anything. That subject will go home with four dollars, and everyone else will go home with 14 dollars. Hence, this experiment models the prisoner's dilemma game, in which the dominant strategy is to defect or not contribute.

What actually happens in these experiments is that some subjects contribute nothing, some contribute the whole 10 dollars, and some contribute a middle amount such as five dollars. Hence, when considering these types of experiments, predictions of the Nash equilibrium are often wrong. Efforts to induce cooperation in these games involve variables such as communication among subjects (face-to-face vs. anonymous text messaging), the number of rounds played, and the extent of players' knowledge about others' payoffs or contributions. One important aspect of cooperation is that a person feel connected to another person (i.e., the scenario is not anonymous).

Another way to look at this problem is to consider sanctions on the free riders. If subjects are willing to sanction free riders, then free riders may be more willing to contribute. The problem is that in order to sanction or punish free riders, a person often has to incur some cost his or herself. Are people willing to impose a cost on themselves to punish free riders? Fehr and Gächter (2000) conducted a repeated public goods experiment using random stranger rematching so that players were, in effect, playing against strangers. The results showed that subjects were willing to punish free riders while imposing costs on themselves.

A revenge emotion might be involved in this game such that a person is willing to take a reduction in payoffs as long as someone else is also taking a reduction.

3. VOLUNTEER DILEMMA

Another type of social dilemma game is referred to as the volunteer's dilemma. Poundstone (1992) reported an experiment initiated in the October 1984 issue of *Science* magazine. In this issue, readers were asked to send in cards requesting 20 or 100 dollars. The researchers promised to provide the requested cash as long as no more than 20 percent of the entries asked for 100 dollars. If more than 20 percent asked for 100 dollars, then no one would get anything. The cooperative strategy would demand that everyone ask for 20 dollars, so that each would receive that amount. However, as in the prisoner's dilemma, there is an incentive for people to request 100 dollars and hope that the thousands of other entries (or at least 80 percent) would only request 20 dollars. Before the results were made public, the magazine grew fearful that too many people would ask for 20 dollars and backed out of the deal. Subsequently, the publication asked readers to pretend that the offer were real. The magazine received 33,511 entries, of which 65 percent requested 20 dollars and 35 percent requested 100 dollars. Although this was not a controlled scientific experiment, it does reveal that if a substantial number of people are willing to cooperate, then defection is the best response strategy; otherwise, cooperating is the best response strategy. This situation illustrates the typical public goods dilemma: when identities are hidden, there is little incentive to cooperate, since a person believes that others will contribute and can receive the benefits of the good without paying the cost.

Governments have figured this problem out and force people to contribute to public goods through taxes. Interest groups and other entities, however, cannot force people to contribute, so they often offer "selective incentives" to motivate people to contribute to a public good. For example, a public radio station may offer an exclusive tote bag for people who donate over 100 dollars to the station. The tote bag might be a status symbol for an individual, since people who listen to public radio will recognize its value. Similarly, the National Rifle Association offers its members magazines, insurance discounts, and a sticker to place on one's car or truck to indicate membership in the organization. In some communities, this is an important status symbol.

While I have argued that prisoner's dilemma and public good games are useful in studying social preferences, Camerer (2003, 46) disagrees:

> [Prisoner's dilemma] and public good games are important in economic life, but they are blunt tools for guiding theories of social preference. These games cannot distinguish between players who are altruistic and players who match expected cooperation. Nor can they distinguish between players who are self-interested and those who have reciprocal preferences but pessimistically think others will free ride.

As mentioned earlier, emotions are very difficult to untangle, but it is not impossible to untangle them or reveal enough of them to get a good measurement. Innovative experimental designs can find ways to isolate single emotions, making these games useful for the study of social preferences.

D. CHICKEN GAMES AND BRINKMANSHIP

1. CHICKEN RUN

In the 1955 movie *Rebel without a Cause,* two teenagers drive stolen cars to a cliff overlooking the ocean to play a game called "chicken run." In this game, two teenagers drive their cars to the edge of the cliff and the first to jump out before the car plunges down the cliff is labeled a chicken. Many variants of this game exist, such as racing a car to beat a train to a crossing. In one popular game theory version conceived by Bertrand Russell (1959), two players drive their cars directly toward each other at a high rate of speed, and the first to swerve away is considered to be the chicken and loses the game. If each player chooses not to swerve, they end up with the worst payoff (−1, −1); if both players swerve, they are both chickens but are at least uninjured (1, 1). However, if one player swerves and the other does not, then the one who does not swerve wins, since the one who swerves is considered to be a chicken (2, 0 or 0, 2). This game is depicted in Table 6.4.

The equilibria to this game is for one player to swerve and the other player to not, and vice versa. However, the best (or Pareto) outcome is for both players to swerve.

One key aspect of this game is the existence of a credible threat or a commitment device (i.e., reputation building). For example, if one of the drivers were to throw his steering wheel out of the car, that action would be considered a credible threat, and the other player should swerve. A player in this game has an advantage if he can make the other player think that he is irrational, crazy, or suicidal.

2. BRINKMANSHIP AND THE CUBAN MISSILE CRISIS

The chicken game is often used to model brinkmanship, which is the art of escalating a dangerous or threatening situation to the point where one side achieves a more advantageous outcome. Brinkmanship is a strategy that is used in labor relations, foreign policy, and international relations. For example, a labor union might call a strike of its workers in the hope that this threat will prompt management to concede to labor demands. In some bargaining situations, parties want to push confrontations to the brink so that the party with whom they are bargaining feels threatened and makes concessions. The North Korean government has used the brinksmanship tactic of testing nuclear weapons in order to receive concessions from Western countries. Terrorist groups use suicide bombings to prompt occupying governments to flee their territories.

The 1962 Cuban missile crisis is the classic example of brinkmanship. This crisis began when the Soviet Union threatened to place nuclear weapons in Cuba in retaliation for the

TABLE 6.4   CHICKEN GAME

		Player 2	
		Swerve	Don't swerve
Player 1	Swerve	1, 1	0, 2
	Don't swerve	2, 0	−1, −1

U.S. placement of missiles in Turkey. Because of Cuba's close proximity to the United States, this action posed a direct threat to American security. In response, President Kennedy used spy planes to discover the existence of missiles, employed a naval blockade of Cuba, and announced an increase in the nation's DEFCON level,[8] indicating that the United States was preparing to invade Cuba. In terms of brinkmanship, this action meant that the United States was escalating the situation. Soon after this information was revealed, President Khrushchev and President Kennedy agreed to remove U.S. missiles from Turkey and Soviet missiles from Cuba.

Consider a game theoretic representation of this game provided by Brams and Kilgour (1988) and presented in Table 6.5. In this case, the United States has two strategies: a naval blockade (B), or an air strike (A) to eliminate missiles already installed. The Soviet Union also has two strategies: withdrawal (W) of their missiles, or maintenance (M) of their missiles.

Like the chicken game, this game has two equilibria: (4, 2) and (2, 4). However, because of U.S. brinkmanship, the outcomes were shifted from the equilibrium position to the compromise positions.

3. HAWK-DOVE GAME

A variant of the chicken game is used widely in biology situations in which animals face conflicts over territory, food sources, mates, and so on. John Maynard Smith and George Price (1973) modeled this conflict as a game that they dubbed the "hawk-dove game." In this game, conflict centers on some divisible resource and there are two strategies that a player can adopt: a Dove strategy—to always retreat if threatened—and a Hawk strategy—to always attack if threatened. If both players use a Hawk strategy, they fight until one is injured, with a 50 percent chance of either being the victor. If both players use a Dove strategy, neither party will be injured and there will be a tie. If one player uses a Dove strategy and the other uses a Hawk strategy, the one who uses a Hawk strategy will win. This game is displayed in Table 6.6.

According to these results, if one player thinks the other will play a Dove strategy, he has an incentive to play a Hawk strategy. In an international relations context, we can imagine two nations disputing ownership of a piece of territory. If they both adopt a Hawk strategy then war will persist, but if they both adopt a Dove strategy then some type of negotiated treaty might prevail. However, war will also prevail if one nation thinks another nation is

TABLE 6.5 US–USSR CHICKEN GAME

		USSR	
		Withdrawal	Maintenance
United States	Blockade	3, 3 (compromise)	2, 4 (Soviet victory)
	Air strike	4, 2 (U.S. victory)	1, 1 (nuclear war)

Source: Reprinted with permission of John Wiley and Sons; Brams and Kilgour 1988.

TABLE 6.6   HAWK-DOVE GAME

		Player 2	
		Hawk	Dove
Player 1	Hawk	0, 0	6, 2
	Dove	2, 6	4, 4

adopting the Dove strategy and consequently adopts the Hawk strategy. In a legal context of two connected businesses disputing an allocation of parking spaces in a shared parking lot, adoption of a Hawk strategy could lead to an expensive legal battle, while adoption of a Dove strategy could lead to a compromise and a reduction of lawyer fees. We can think of a hawk-dove game as a game in which both parties have a vested interest to not escalate into the (Hawk, Hawk) cell and to cooperate with each other to pull themselves into the (Dove, Dove) cell. But the assumption of rationality diverts this outcome, since a player who knows that the opponent will play Dove has an incentive to play Hawk.

4. ACME-BOLT TRUCK EXPERIMENT

In an attempt to examine chicken games and other games in which direct conflict leads to disastrous outcomes, the Acme-Bolt experiments were devised in the early 1960s by two social psychologists (Deutsch and Krauss 1960). The premise of this game is that two trucking companies, one called Acme and the other called Bolt, are paid based on the load they are carrying and, more importantly, on how much the trip costs. As a result, both trucking companies have an incentive to take the shortest routes to their drop-off location, since this strategy will save valuable gas money and produce lower costs. Drivers can take two routes: one is a long route that will cost more in terms of gas mileage, and the other is a short route that will save on gas mileage. Figure 6.1 shows the two routes, with the shorter route being the one-lane middle road. The dilemma is that since the shorter route is a one-lane highway, if both drivers decide to take that route, "deadlock" will occur and one of the trucks will have to turn around to let the other truck through. Consequently, it is in each truck's best interest to get through the one-lane road first, since the one who backs down will earn less money. Both trucking companies are better off if they come to an agreement over the use of the one-lane road.

Deutsch and Lewicki (1970) altered the basic game to make it resemble a chicken game. In this treatment, subjects were notified that if the two trucks met at any point on the one-lane highway, this event would be considered a collision and the experiment would terminate. Second, they introduced a commitment device in the form of a locking device, which, once installed, committed the truck to moving forward (i.e., there would be no way to stop the truck). When one player used a lock, the other player was notified. Deutsch and Lewicki conducted three treatments with different locks: bilateral locks (both players had locks), unilateral lock (only an Acme player had a lock), and no locks for either player.

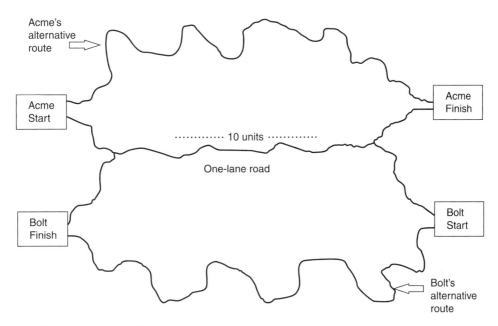

FIGURE 6.1 Bolt–Acme dilemma

Source: Reprinted with permission of SAGE Publications; Deutsch and Lewicki 1970.

Players were given an initial endowment of two dollars (remember that this experiment was conducted in the 1970s) and were paid 60 cents to make the journey. However, they incurred an operating fee of one cent per minute. During the experiment subjects were placed in a cubicle, where they had control of a toy truck and could see the experimenter but not the other subject. The researchers stressed that subjects should not worry about their opponent, but should just try to make as much money as possible for themselves.

The researchers then varied the number of times the stage game was played. In one treatment, subjects played a one-shot staged game, in which the Acme players who had a lock were the most successful in terms of payoffs. In another treatment, the stage game was repeated for 20 rounds, and whoever had a locking device was the most successful in terms of payoffs. Essentially, these experiments show that having a commitment device does lend an advantage to the player who owns the device. Deutsch and Lewicki (1970, 377) conclude:

> Perhaps all of this can be summed up by saying that "locking oneself in" to an irreversible position in order to gain an advantage is rarely more beneficial than cooperating with the other for mutual gain, and it has the prospect of becoming a mutually destructive contest of wills.

E. BATTLE OF THE SEXES GAME AND COORDINATION GAMES

1. CLASSIC STORY OF THE BATTLE OF THE SEXES

The classic (and unfortunately sexist) "battle of the sexes" game depicts a husband and wife deciding whether to go to a ballet or a boxing match (Luce and Raiffa 1957). The husband most prefers to go to a boxing match while the wife most prefers to go to the ballet. Each prefers to be together rather than alone, but the husband gets more utility if they go to the boxing match (1, 2) and the wife gets more utility if they go to the ballet (2, 1). Each spouse receives his or her lowest utility if he or she goes alone (0, 0). This game has two pure strategy equilibria in which the husband and wife both go to the boxing match or both go to the ballet.[9] Therefore, both players have an incentive to cooperate, but the problem is coordination, since they do not agree on which outcome is better. In this game, bargaining may lead to a solution or the wife could restrict alternatives and say that she faints at the sight of blood, forcing her husband to go to the ballet. The normal form game is presented in Table 6.7.

(Ballet, Ballet) and (Fight, Fight) are both equilibria, so which equilibrium will prevail if some outcome is demanded? That question is the heart of coordination problems. What do you think would happen if we assumed that one player observed another player's strategy choice before selecting her own strategy choice? It could solve the game, but we will talk about sequential games in later chapters.

2. COORDINATION IN A MATCHING PENNIES GAME

In games with multiple equilibria, coordination can be a problem that players might not be able to overcome. Consider the following matching pennies game in Table 6.8.

In this game, Player 1 wants to play Up when Player 2 plays Left and play Down when Player 2 plays Right. The question is: How can each player determine what the other player will play? This coordination problem can be easily overcome with communication. For example, consider an experiment in which two subjects are playing this game against each

TABLE 6.7   BATTLE OF THE SEXES GAME

		Husband	
		Ballet	Fight
Wife	Ballet	2, 1	0, 0
	Fight	0, 0	1, 2

TABLE 6.8   MATCHING PENNIES GAME

		Player 2	
		Left	Right
Player 1	Up	1, 1	0, 0
	Down	0, 0	1, 1

other and are seated back to back in the laboratory so that they cannot see how the other will play. But the experimenter is distracted for a moment, and the first subject taps the other subject on his back so that he turns around, and, without saying a word, the first subject raises his thumb in the up position. This cheating maneuver would be enough to signal that (Up, Left) is the strategy to play. However, without cheating, the two subjects might simply think that it is reasonable to select their first strategies, so Player 1 selects Up and Player 2 selects Left.

3. FOCAL POINT EQUILIBRIUM

A focal point is a psychologically prominent outcome that is used in the absence of communication and seems to be a natural outcome to a game (Schelling 1960). Schelling (1960, 57) comments:

> People *can* often concert their intentions or expectations with others if each knows that the other is trying to do the same. Most situations—perhaps every situation for people who are practiced at this type of game—provide some clue for coordinating behavior, some focal point for each person's expectation of what the other expects him to expect to be expected to do.

This focal point is often referred to as the Schelling point. To illustrate this idea, Schelling (1960, 56–57) provided the following hypothetical questions (of which I only provide a sample)[10]:

1. Name heads or tails. If you and your partner name the same, you both win a prize.
2. Circle one of the numbers listed in the line below. You win if you all succeed in circling the same number:

   7 100 13 261 99 555

3. You are to meet somebody in New York City. You have not been instructed where to meet; you have no prior understanding with the person of where to meet, and you cannot communicate with each other. You are simply told you have to guess where to meet and that he is being told the same thing, and that you will just have to try to make your guesses coincide.
4. Name any amount of money. If you all name the same amount, you can have as much as you named.
5. You are to divide $100 into two piles, labeled A and B. Your partner is to divide another $100 into two piles labeled A and B. If you allot the same amounts to A and B, respectively, that your partner does, each of you gets $100 dollars; if your amounts differ from his, neither of you get anything.

Although there are no correct answers to these questions, it is expected that reasonable people should be able to coordinate on the same answers.[11] When there are multiple equilibria in game theory, it is possible that something about the game's context or environment leads players to one equilibrium over another. For instance, assume that the only difference between two environments is a tiny scratch in one of them. If players could locate this scratch, it could serve as a focal point that leads them to a single equilibrium.

F. STAG HUNT OR ASSURANCE GAMES

1. THE ROUSSEAU GAME AND RISK-DOMINANT EQUILIBRIUM

In his *Discourse on Inequality*, Rousseau (1755) posed the following problem. Assume that two hunters are starving for food and decide to hunt a stag for dinner. This endeavor requires coordination, since it takes more than one hunter to actually catch the stag. If both hunters remain at their posts, they will catch a stag and both will be well fed, receiving a payoff of (3, 3). Now, assume that you are at your post and a rabbit passes by. Since it only takes one hunter to catch a rabbit, if you defect and decide to hunt the rabbit, you will catch it and be well fed, but the meal will not be as satisfying as the stag. Plus, you will have eaten while the other hunter will not, so you will receive a payoff of 2 and the other will receive a payoff of 0. If both hunters defect, then they will decrease their probability of catching the rabbit, since there are two hunters and only one rabbit, reducing payoffs to (1, 1). This coordination game is portrayed in Table 6.9.

This game has two equilibria: one when the players trust each other and hunt a stag and the other when they each take a risk and hunt the rabbit. Notice that the Hunt Stag equilibrium is the Pareto dominant equilibrium, since this strategy is better for both players. The Chase Rabbit equilibrium has the highest guaranteed payoff for both players. That is, no matter what the other player does, by selecting Chase Rabbit, a player guarantees himself at least a payoff of 1. The Chase Rabbit equilibrium is called a risk-dominant equilibrium, since it does not depend on whether other players defect. The temptation to defect arises only when you believe that others will defect. These types of games are referred to as assurance games or trust games, because a player needs assurance that the other players will not defect (and chase a rabbit). If a player believes that others will defect, then she has an incentive to defect also.

2. QUORUM BUSTING

To illustrate a stag hunt game in the real world, consider a legislative strategy referred to as quorum busting, in which a minority party can attempt to defeat legislation favored by a majority party. Most legislatures have a quorum requirement that sets the minimum number of members needed to conduct official business. If not enough members are physically present in the legislature, then a quorum requirement is not fulfilled and no official business, such as voting on bills, can be conducted. To compel attendance, a presiding officer can issue a "call of house," which legally orders members to attend a deliberation and can sanction members financially if they do not comply.[12] Consequently, minority members must consider costs when deciding whether to join a quorum-busting coalition. If enough

TABLE 6.9 STAG HUNT GAME

		Player 2	
		Hunt stag	Chase rabbit
Player 1	Hunt stag	3, 3	0, 2
	Chase rabbit	2, 0	1, 1

members of the minority party hide to keep a quorum from being established, then it is in the interest of other minority party members to join them. In this case, the possible costs are absorbed, since the members' collective efforts will have stopped voting on a piece of legislation the group opposes. But if not enough minority members plan to hide, then it is not in the interest of other minority members to join them, since the members will only absorb the cost and not the benefits of halting voting.[13] Hence, like the stag hunt game, quorum busting achieves the Pareto optimal outcome when members trust each other and join the coalition.[14]

3. EXPERIMENT ON STAG HUNT GAME: COMMUNICATION AND TRUST

To coordinate and achieve the Pareto optimal outcome in stag hunt games, players have to build a system of trust. Cooper et al. (1994) conducted a laboratory stag hunt experiment in which subjects were able to communicate with each other by using cheap talk. To understand this term, first assume that there are two players and one player sends a message about some aspects of a player's payoff to another player. If there are no consequences to a player for sending the message (i.e., there are no payoff-specific costs involved), then the message has no effect on her. It doesn't matter to her if she sends the message, but if she sends it, it could impact the receiver's payoffs. For example, a person might send a message that is a lie, which is costless for the sender but could change the receiver's behavior if she believes it.

Two cheap-talk treatments were conducted: a one-way communication treatment in which either Player 1 or Player 2 was able to send a message to the opponent, and a two-sided communication treatment in which both players were able to send a message to the other. The message simply instructed a player which strategy another player planned to play.

The experiments were conducted in a computer lab where three subjects played a series of two-person games. The subjects played in a series of rounds in which they were randomly matched to different players so that they were unaware of whom they were playing against. Depending on the treatment, subjects played for 11 to 21 periods. The computers each displayed a payoff matrix, as shown in Table 6.10. This matrix is similar to the matrix in Table 6.9, but it is inverted so that (2, 2) is the dominant preferred

TABLE 6.10 STAG HUNT PAYOFFS

		Player 2	
		1	2
Player 1	1	800, 800	800, 0
	2	0, 800	1000, 1000

Source: Reprinted with permission of Oxford University Press and the President and Fellows of Harvard College; Cooper et al. 1992.

outcome. The values are units of probabilities (where one thousand points guaranteed a win of one dollar).[15]

I will discuss three treatments here, although more were conducted, and examine only the last 11 rounds of the experiment. First, when this game was played without any treatments (i.e., only the raw game), then the outcome of (1,1) prevailed 97 percent of the time. There was essentially no trust among subjects in this treatment. Interestingly, in the one-sided communication treatment, 53 percent of the subjects selected (2, 2) 52 percent of the time—a vast improvement over the previous results in terms of trust. However, when two-way communication was allowed, the (2, 2) outcome was selected 91 percent of the time. Consequently, when both players were able to establish communication, then trust increased. Cooper et al. (1994, 142) comments:

> In the coordination games, however, there is no question about the desired outcome. The problem here, given the strategic uncertainty and the presence of a safer strategy does not support the best outcome, is for players to develop sufficient confidence that their rivals will play the strategy which supports the Pareto-dominated equilibrium. Among the institutions studied, this is best accomplished by allowing both players an opportunity to announce their intentions.

Hence, when both players are able to communicate with each other, trust is established, but when only one player is allowed to communicate, trust is only vaguely established. Again, some psychological principal governs this exchange that is not explained by rational choice theory. Two-way communication, not one-way communication, appears to be a coordination device that enables coordinated action.

4. COORDINATION AND ELITES

Eckel and Wilson (2007) examined a coordination game similar to the game used by Cooper et al. (1994) but incorporated a hierarchical structure in which a higher status player (a royal family) could give a coordination signal, but a "commoner," or lower class player, could not. To differentiate between low- and high-status players, subjects had to take a quiz prior to the experiment in which they were paid for correct answers. Subjects who received high scores were considered part of the royal family and subjects who received low scores were classified as commoners. Subjects knew the move that the opposite subjects had made prior to making their own move. The question was whether subjects in the experiment would follow the lead of high-status subjects as opposed to low-status subjects. The experiment did indeed find that subjects followed the lead of high-status subjects more often, indicating that socially influential leaders might be able to direct coordination problems in their favor.

G. SUMMARY

In this chapter, I presented four classic game theory models and showed how their "unique" results generated a fruitful literature. First, I illustrated the repeated prisoner's dilemma game, which in the one-shot version has an equilibrium of both players defecting, but which can achieve cooperation for sustained periods in repeated trials. However, more stringent tests of repeated games have found less cooperation. Next, I examined the chicken game, which is used to model deterrence, and showed that players who possess a credible threat have an advantage in these games. Examples of coordination games and experiments illustrated how games with multiple equilibria might be solved by using a focal equilibrium, or how reasonable conditions within the model might trigger an equilibrium. Finally, I examined assurance games, in which equilibrium is dependent on the trust of other players. Experiments showed that trust is only ensured if both players can communicate with each other. Large bodies of literature exist for all of these. The interested reader looking for more experimental examples should consult Roth (1995a and 1995b), Camerer (2003), and Morton and Williams (2010). Subsequent chapters will revisit issues raised in this chapter.

NOTES

1. In other repeated games, payoffs can vary from round to round when they are additive across periods.
2. Finite games have a known ending period, whereas the ending period of an infinite horizon game is unknown. Knowledge of this ending constraint can bias players' behavior. In experiments that conduct repeated trials, either subjects know the exact ending constraints, or there is some sort of probabilistic ending constraint so that subjects do not know the exact number of periods in which they will participate, but they do know when the end is near.
3. See Axelrod (1984) for examples of these types of strategies in real-world situations.
4. A repeated prisoner's dilemma game can be played at http://www.gametheory.net/Mike/applets/PDilemma.
5. He also reports on an additional tournament that included more competitors. For this tournament, the tit-for-tat strategy again performed well.
6. This game is similar to the El Farol bar dilemma presented in Problem 6.7.
7. The authors conducted many more treatments. I report on only one here.
8. DEFCON status indicates the defense readiness condition, with DEFCON 1 representing imminent attack.
9. In the next chapter, we will solve for the mixed strategy equilibrium of this game.
10. For an alternative but similar experimental design, see Mehta, Starmer, and Sugden 1994.
11. The answers that received a majority were: (1) heads; (2) 7; (3) Grand Central Station information booth at noon; (4) one million; (5) 50/50.
12. See Palmer (2006) for more information about quorum procedures.
13. For example, in 2002, Texas Democrats in the state legislature hid to avoid a vote on a redistricting plan, and in 2011, Wisconsin Democrats in the state legislature hid to avoid a vote on a labor plan.
14. Also see Dasgupta et al. (2008) for a stag hunt game represented by a sequential majority rule voting game.
15. This procedure is used to ensure that subjects have risk-neutral preferences in regard to maximizing expected utility.

SOLVING FOR MIXED STRATEGY EQUILIBRIUM

A. ROCK, PAPER, SCISSORS

The Rock, Paper, Scissors (RPS) Annual Championships are a serious competition for "the professionals" who participate in this tournament. The following is an excerpt from the official RPS website, which outlines a guide for advanced techniques.

The basic skills of RPS need no discussion. Most children can be taught to form the three throws with their hands and with a little practice can follow the prime and reveal their chosen throw at the appropriate time.

An advanced RPS player can do more than that. He can use his hands to confuse or deceive an opponent. She can make her opponent believe she is going to throw Rock when she is actually going to throw scissors.

"Cloaking" is the term used for delaying the unveiling of the throw. Put a little more simply, "Cloaking" is waiting until the last possible second to throw Paper or Scissors....

Another step beyond cloaking, "shadowing" is pretending to throw one thing, but changing to another at the last possible moment. This is much more difficult and requires great care in execution....

"Tells" are visible behaviors through which a player may unconsciously reveal a throw to an opponent....The face and lips are common places to find tells....Serious RPS players will spend time hunting for their own tells...and learning to suppress them....

Proponents of the "Chaos School" of RPS try to select a throw randomly. An opponent cannot know what you do not know yourself. In theory, the only way to defeat a random throw is with another random throw—and then only thirty-three percent of the time. Critics of this strategy insist that there is no such thing as a random throw. Human beings will always use some impulse or inclination to choose a throw, and will therefore settle into unconscious but nonetheless predicable patterns....

The use of Gambits in competitive RPS has been one of the greatest and most enduring breakthroughs in RPS strategy. A "Gambit" is a series of three throws used with strategic intent. "Strategic intent" in this case, means that the three throws are selected beforehand as part of a planned sequence. Selecting throws in advance helps prevent unconscious patterns from forming and can sometimes reduce tells. (World RPS Society 2009)

While we might all snicker at the thought that competitive RPS entails skill, the fact is that selecting strategies randomly is a skill that many of us do not possess. The points raised by the advanced RPS techniques are relevant to some game theory models in which the concern is for how you play a strategy against your opponent with the intent of concealing your strategy from him or her. Consider the ace/king example in Chapter 1. In this game, you must place an ace and a king randomly so that your opponent cannot figure out the order in which you placed them. If this game is played repeatedly and your order is not random, then over time your opponent will have an easier time guessing the placement of the cards.[1] In this chapter I will present other games in which the importance of randomness is relevant.

The excerpt on competitive RPS also reveals that this game would cease to be a competition if a player could use a random generating device to make her decisions. What makes the game competitive is that players must generate random strategies in their minds. However, experimental studies have shown that laboratory subjects have a difficult time making random selections (see Binmore, Swierzbinski, and Proulx 2001).[2] Consequently, many experiments that incorporate "randomness" into their design have a computer make random selections for the subjects. A subject in an experiment might know that she wants to randomize three strategies 33 percent of the time but may be unable to do this in repeated trials. To help subjects with this task, experimenters provide subjects with a random generating device such as a computerized urn. A program generates an urn with 99 marbles, of which 33 are marked with an R, 33 are marked with a P, and 33 are marked with a S. During game play, the program simply selects a marble from the urn, which in turn produces random responses. Without the aid of a random generating device, advanced RPS skills are necessary to become a champion RPS player.

In this chapter we will examine the concept of a mixed strategy, which is the same idea as randomly selecting marbles from an urn. Mixed strategies are a way to provide a Nash equilibrium solution to constant sum and variable sum games. In the next section I introduce the notion of mixed strategies and explain how to calculate them. In the following section I explain how mixed strategies are interpreted and then present experiments that test mixed strategy equilibrium. In the last section I present some observational results on mixed strategies.

B. CALCULATING MIXED STRATEGIES

1. SPADES-HEARTS GAME

Mixed strategies are used when two or more players are attempting to randomize the strategies they intend to implement. In some cases, the players want to outguess their opponent and so selecting an optimal randomization mix such as 33 percent will garner optimal long-term payoffs, and in other cases, players must randomize strategies but would like their opponent to correctly "guess" what strategy they are playing (such as in a coordination game). Calculating mixed strategies for each player in the game allows probabilities to be assigned to the selection of strategies, which in turn provides a solution. By selecting a mixed strategy, players are attempting to make each other indifferent over which strategy is selected. Morton and Williams (2010, 235) comment:

Theoretically, when an individual chooses a mixed strategy, she is mixing between pure strategies such that the other players in the game are indifferent over their own strategies and vice versa. Actors do not mix necessarily because it is their best response; they mix because they are indifferent and mixing makes other actors indifferent as well.

Let us consider the use of mixed strategies in a game of pure conflict, as depicted by the spades-hearts card game. Assume that two players each have a heart and a spade. The players lay both cards facedown so that their identity is hidden. Each player is allowed to reveal one card. If Player 1 reveals a spade and Player 2 reveals a heart, then Player 1 loses one dollar and Player 2 gains one dollar. If Players 1 and 2 both reveal a heart, then Player 1 loses four dollars and Player 2 gains four dollars. If Player 1 reveals a heart and Player 2 reveals a spade, then Player 1 gains three dollars and Player 2 loses three dollars. Finally, if Player 1 plays a spade and Player 2 plays a heart, then Player 1 gains two dollars and Player 2 loses two dollars.

In this game, Player 2 is at a disadvantage. Consider that what Player 1 wants is either to play a heart when Player 2 plays a spade, or to play a spade when Player 2 plays a heart. Player 2 wants either to play a heart when Player 1 plays a heart, or to play a spade when Player 1 plays a spade. This game is depicted in Table 7.1.

Notice that there is no pure strategy equilibrium. Neither knows what the other will do, but we can calculate a strategy that will guarantee the players their maximum payoff in the long run. Notice that if Player 2 just assumed a 50/50 mix from Player 1 and played spades and hearts in a 50/50 mix, then he would expect to get $0.5(1) + 0.5(-3) = -1$ from playing spades and $0.5(-2) + 0.5(4) = 1$ from playing hearts. His overall expected utility would be 0. If Player 1 assumed that Player 2 was playing a 50/50 mix and played her cards in a 50/50 mix, she would expect $0.5(-1) + 0.5(2) = 0.5$ from playing spades and $0.5(3) + 0.5(-4) = -0.5$ from playing hearts, for an expected utility of 0. However, notice that Player 2 has an incentive to play hearts more often than Player 1, since his expected payoff from playing hearts is 1, whereas Player 1's expected payoff is -0.5. Therefore, Player 1 needs to anticipate that Player 2 will play hearts more than she will. Similarly, Player 2 needs to anticipate that Player 1 will play spades more often than he will, because her expected payoff from spades is higher.

Hence, the problem is that both players want to use one strategy more often than another, but they still want to maintain the element of uncertainty. One way to randomize is to use an urn that contains 100 red and black marbles. If a player selects a red marble he plays hearts, and if he selects a black marble, he plays spades. The question then becomes: What should the distribution of black and red marbles in the urn be for each player? This question can be answered by solving for the mixed strategy of both players. A mixed strategy

TABLE 7.1   SPADES-HEARTS GAME

		Player 2	
		Spades	Hearts
Player 1	Spades	−1, 1	2, −2
	Hearts	3, −3	−4, 4

is a probability distribution over a set of strategies (in our case, whether to play hearts or spades). This distribution allows us to calculate the expected utility for a particular mix of strategies. We can then define a mixed strategy equilibrium (MSE), in which each player uses the best mixed strategy (or best response) against his or her opponent's best mixed strategy (or best response).

2. MIXED STRATEGY EQUILIBRIUM FOR SPADES-HEARTS GAME

Before we learn to calculate mixed strategies and specify an MSE, it is helpful to discuss the logic behind calculating this type of strategy set. Recall that a Nash equilibrium occurs when no player has a positive incentive to defect from a cell. That is, the players are in what Osborne (2004) calls a "steady state." He notes (99):

> In a steady state, every player's behavior is the same whenever she plays the game, and no player wishes to change her behavior, knowing (from her experience) the other player. In a steady state in which each player's "behavior" is simply an action and within each population all players choose the same action, the outcome of every play of the game is the same Nash equilibrium.

This steady state is our goal in calculating mixed strategies to specify an MSE. When we calculate mixed strategies, we want to make players indifferent between the strategies that they are randomizing. A player wants to keep the opponent guessing, and calculating each player's optimal mixed strategies guarantees that players will not know what their opponent will play with certainty. When calculating an MSE, the probabilities used by a player to calculate her mixed strategy are based not on her own payoffs, but on the payoffs of the other player. This is because the players know each other's payoffs, and so a player needs to know what the other player will do. Calculating each other's strategies so that each player is indifferent to the mix they are playing constitutes a best response to each mixed strategy that is played. This indifference produces a steady state.

In our spades-hearts game, a mixed strategy for Player 1 is the probability distribution $(p, 1 - p)$, where $0 \leq p \leq 1$, and a mixed strategy for Player 2 is $(q, 1 - q)$, where $0 \leq q \leq 1$. Note that for Player 1 the strategy $(1, 0)$ is the pure strategy of playing spades every time, and the strategy $(0, 1)$ is the pure strategy of playing hearts every time. We must now calculate the players' best response strategy to each other. Assume that Player 1 thinks that Player 2 will play spades with probability q and hearts with probability $(1 - q)$. Player 1's beliefs about how often Player 2 will play spades are outlined in Table 7.2.

To calculate q for Player 1, we must examine her outcomes when she plays spades (-1 and 2) and set them equal to her outcomes when she plays hearts (3 and -4). So, given q, the expected payoff to Player 1 from choosing spades is $-1(q) + 2(1 - q)$, or $2 - 3q$. Given q, the expected payoff to Player 1 from choosing hearts is $3(q) + -4(1 - q)$, or $7q - 4$. Notice that when $q = 0$, then $2 - 3q > 7q - 4$ and she should play spades, and when $q = 1$, then $2 - 3q + 2 < 7q - 4$ and she should play hearts. Player 1 is indifferent when $2 - 3q + 2 = 7q - 4$, or when $q = 0.6$.

TABLE 7.2 SPADES-HEARTS GAME WITH PROBABILITIES

		Player 2	
		Spades (q)	Hearts (1–q)
Player 1	Spades (p)	–1, 1	2, –2
	Hearts (1 – p)	3, –3	–4, 4

How should Player 1 behave? For a given q, she should pick the action that maximizes her expected payoff. This action is denoted by her best response function $B_1(q)$, which is the set of probabilities that Player 1 assigns to spades in response to q, where $\{0 \leq p \leq 1\}$ is the mix Player 1 is playing. Player 1's $B_1(q)$ is:

$$B_1(q) = \begin{cases} \{1\} & \text{if } q < 0.60 \\ \{p: 0 \leq p \leq 1\} & \text{if } q = 0.60 \\ \{0\} & \text{if } q > 0.60 \end{cases}$$

When q < 0.6, Player 1 should play spades (with q = 0 meaning that spades should be played with certainty), and when q > 0.6, she should play hearts (with q = 1 meaning that hearts should be played with certainty). Consider Figure 7.1, which graphically represents $B_1(q)$ for Player 1.

The horizontal axis presents the probability that Player 1 will play spades in a mixed strategy, and the vertical axis presents the probability that Player 2 will play spades in a mixed strategy. Note that the horizontal axis represents Player 1's pure strategy of playing a spade and zero represents her pure strategy of playing a heart. At 60 percent she is indifferent between playing spades and hearts.

To calculate p for Player 2, we must examine his payoffs when he plays spades (1 and –3) and set that equal to his payoffs when he plays hearts (–2 and 4). Given p, the expected payoff to Player 2 from choosing spades is $1(p) + -3(1 - p)$, or $4p - 3$. Given p, the expected payoff to Player 2 from choosing hearts is $-2(p) + 4(1 - p)$, or $4 - 6p$. Again, notice that when p = 1, then $4p - 3 > 4 - 6p$ and he should play spades, and when p = 0, then $4p - 3 < 4 - 6p$ and he should play hearts. He is indifferent when $4p - 3 = 4 - 6p$, or when p = 0.7.

Now we can calculate $B_1(p)$ for Player 2, which is:

$$B_1(q) = \begin{cases} \{1\} & \text{if } p > 0.70 \\ \{q: 0 \leq q \leq 1\} & \text{if } p = 0.70 \\ \{0\} & \text{if } p < 0.70 \end{cases}$$

When p > 0.7, Player 2 plays spades (with p = 1 meaning that spades should be played with certainty), and when p < 0.7, he plays hearts (with p = 0 meaning that hearts should be played with certainty). Player 2's $B_1(p)$ is graphically presented in Figure 7.2.

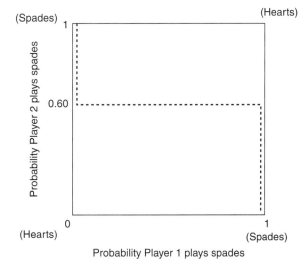

FIGURE 7.1 $B_1(q)$ for Player 1

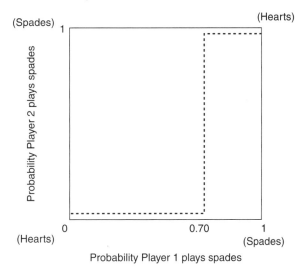

FIGURE 7.2 $B_1(p)$ for Player 2

Figure 7.3 contains the best response functions for Players 1 and 2 that were depicted in the prior two figures. Recalling our discussion of Nash equilibrium and steady state, note that this state is achieved when both players' best response functions intersect at 0.60 and 0.70.

Finally, we can calculate the value or expected utility of the game, which is how much each player is guaranteed if they play these best response strategies. Since the best response strategy depends on the actions of the other player, we must use Player 1's beliefs about Player 2's actions to calculate expected utility for Player 1. If we fix Player 1's beliefs at (0.6, 0.4), then her expected utility from playing spades is $EU_1(S) = 0.6(-1) + 0.4(2)$, or 0.2. Similarly, if we fix Player 1's beliefs at (0.6, 0.4), then her expected utility from playing hearts is $EU_1(H) = 0.6(3) + 0.4(-4)$, or 0.2.

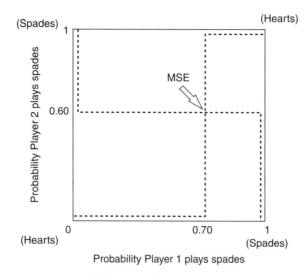

FIGURE 7.3 MSE for spades-hearts game

In this case, if Player 2 selects the probabilities (0.6, 0.4), then Player 1 has no positive incentive to switch from one strategy to another. The same calculation can be made for Player 2. If we fix Player 2's beliefs at (0.7, 0.3), her expected utility from playing spades is $EU_2(S) = 0.7(1) + 0.3(-3)$, or -0.2. If we fix her beliefs at (0.7, 0.3), then her expected utility from playing hearts is $EU_2(H) = 0.7(-2) + 0.3(4)$, or -0.2.

Again, if Player 1 selects the probabilities (0.7, 0.3), then Player 2 has no positive incentive to shift from one mixed strategy to another. These beliefs establish that Player 1's mixed strategy is a best response to Player 2's mixed strategy (and visa versa). An equilibrium to this game can be written as MSE = (p = 0.7, q = 0.6 | 0.2, −0.2). Notice that Player 2's expected utility is −0.2, establishing his disadvantage in this game. Note that because this is a zero-sum game, the players' expected payoffs are the exact opposite of each other. By calculating the mixed strategy equilibrium, players are expected to receive their optimal payoff, given the constraints of the game, and should have no incentive to deviate from this course of action. We can think of mixed strategies as being a weighted average of a game's pure strategies, so that mixed strategies represent both players' best response strategies and players are indifferent between strategies in equilibrium.

3. WHY WOULD A PLAYER USE A MIXED STRATEGY?

Now that we have learned how to calculate mixed strategies for players, several questions remain: How useful is this calculation? What does it mean for players to play mixed strategies? Thus far, we have only been concerned with the classical version of mixed strategies in which players try to conceal the strategy they are attempting to play. Von Neumann and Morgenstern (1944, 146) note:

Thus one important consideration for a player in such a game is to protect himself against having his intentions found out by his opponent. Playing several such strategies at random, so that only their probabilities are determined, is a very effective way to achieve a degree of such protection: By this device the opponent cannot possibly find out what the player's strategy is going to be, since the player does not know it himself.

The problem with this view of mixed strategies is that it applies to games of conflict such as zero-sum games. In many non-zero-sum games, players have a common interest and do not want to conceal their strategy choice; on the contrary, they want their strategy choices to be known. Schelling notes that "in games that mix conflict with common interest…randomization plays no such central role" (1960, 175) as in our second example, the battle of the sexes game, in which players want their strategy choices to be transparent.

In games in which players have a common interest, mixed strategies can represent uncertainty about the strategy each player will play. Gibbons (1992, 152) remarks:

> Put more evocatively, the crucial feature of a mixed-strategy Nash equilibrium is not that player j chooses a strategy randomly, but rather that player i is uncertain about player j's choice; this uncertainty can arise either because of randomization or (more plausibly) because of a little incomplete information.

Additionally, Morrow (1994, 87) notes:

> A mixed strategy…does not give random chances of the pure strategies. Rather, it delineates the uncertainty in the mind of the other player about the mixing player's strategy. The other player's indifference between its strategies captures the degree of uncertainty where the other player cannot exploit the mixing player.

The next section presents a solution to a coordination game in which mixed strategies represent uncertainty about a player's choice.

4. MIXED STRATEGY EQUILIBRIUM FOR THE BATTLE OF THE SEXES GAME

Reconsider the battle of the sexes game discussed in the last chapter. Recall that this game has two equilibria (Ballet, Ballet) and (Fight, Fight)—as well as an MSE (see Table 6.7).

To calculate the MSE for this game, let us calculate Player 1's expected payoffs when Player 2 selects Ballet and when he selects Fight (and vice versa for Player 2).

Given q, the expected payoff to Player 1 when Player 2 selects Ballet is $2(q) + 0(1 - q)$, or $2q$.

Her expected payoff when Player 2 selects Fight is $0(q) + 1(1 - q)$, or $1 - q$. When $q = 1$, $2q > 1 - q$ and Ballet should be selected, and when $q = 0$, $2q < 1 - q$ and Fight should be selected. Player 1 is indifferent when $2q = 1 - q$, or $q = 1/3$.

Given p, the expected payoff to Player 2 when Player 1 selects Ballet is $1(p) + 0(1 - p)$, or p.

His expected payoff when Player 1 selects Fight is $0(p) + 2(1 - p)$, or $2 - 2p$. When $p = 1$, $p > 2 - 2p$ and Ballet should be selected, and when $p = 0$, $p < 2 - 2p$ and Fight should be selected. He is indifferent when $p = 2 - 2p$, or $p = 2/3$.

Figure 7.4 shows the best response functions for Players 1 and 2.

TABLE 7.3 BATTLE OF THE SEXES REVISITED

		Player 2 (Husband)	
		Ballet	Fight
Player 1 (Wife)	Ballet	2, 1	0, 0
	Fight	0, 0	1, 2

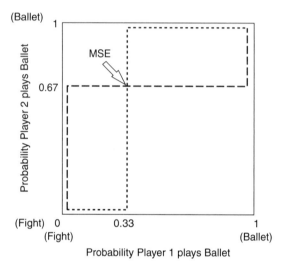

FIGURE 7.4   MSE for battle of the sexes game

Notice that the best response functions meet at three points. Two are the (Fight, Fight; q = 0, p = 0) and (Ballet, Ballet; q = 1, p = 1) points, which are both pure strategy equilibria. The final point is the mixed strategy equilibrium of (q = 1/3, p = 2/3, 0.6, 0.6).[3]

C. EXPERIMENTAL TESTS OF MIXED STRATEGY EQUILIBRIUM

1. O'NEILL'S EXPERIMENT

Recall that in the last chapter, payoffs were essentially the value presented in the normal form when players selected pure strategies. However, the introduction of mixed strategies presents a problem, since players essentially have preferences over lotteries and risk behavior might bias outcomes. A researcher who wants to model mixed strategies wants to have mixed strategy solutions that are invariant to the way that subjects transform dollar payoffs into expected utilities. Remember the prior discussion of risk aversion in which an exponential or square root function translated preferences into risk-loving and risk-hating behavior. To cancel out risk-loving and risk-hating behavior so that behavior can be modeled as risk neutral, as we discussed in Chapter 4, binary lotteries can be used to eliminate risk. When this method for controlling risk behavior is used in the laboratory, it is called explicit randomization. This term means that subjects are given an explicit device that

TABLE 7.4   O'NEILL'S GAME (1987)

		Column					
		Ace	2	3	Joker	MSE	Actual
	Ace	5	−5	−5	−5	0.2	0.221
	2	−5	−5	5	5	0.2	0.215
	3	−5	5	−5	5	0.2	0.203
Row	Joker	−5	5	5	−5	0.4	0.362
	MSE	0.2	0.2	0.2	0.04		
	Actual	0.226	0.179	0.169	0.426		

Source: O'Neill 1987.

helps them to randomize their choices. At the beginning of this chapter, we discussed the problems that people have with generating random actions. Consequently, researchers give subjects a random device like an urn filled with marbles to help them randomize choices. This method controls for risk aversion, since it offers subjects a choice of lotteries: one that risk lovers prefer and one that risk haters prefer. When these choices are randomly distributed through explicit randomization, risk behavior can be controlled for. In this section I will present experiments that use explicit randomization to study games that have mixed strategy solutions.

O'Neill (1987) studied mixed strategy equilibrium by presenting a card game in which two players are dealt four cards each: an ace, a two, a three, and a joker. The players lay the four cards facedown in front of them and each player reveals the identity of one card. Player 1 wins if two jokers are played or the number cards do not match. Player 2 wins if the number cards match or the joker is not matched. Hence, explicit randomization is achieved in this game by having subjects play mixed strategies by selecting cards. In O'Neill's experiment, players were endowed with $2.50 in nickels. The game is presented in Table 7.4.

Notice that there are only two types of payoffs, 5 and −5 (ignore the rightmost two columns and bottom two rows for the moment). In this game, risk aversion plays no role, since there are only two possible outcomes. So although there is a mixed strategy solution, subjects are essentially playing a pure strategy game. In Table 7.4, the mixed strategy equilibrium is denoted in the MSE column, where each player should play the vector (0.2, 0.2, 0.2, 0.4), and when the game is normalized, the expected utility is 0.4 for the row player and −0.4 for the column player.

The subjects played the game 105 times using their stash of nickels. The actual outcomes are listed in the last column and last row of the table. As specified, Player 1 played a joker 0.362 of the time and Player 2 played a joker 0.426 of the time, which are close to the predicted outcomes.

O' Neill's experiment was criticized on technical grounds because subjects' choices were serially correlated and not random (see Brown and Rosenthal 1990). Another experiment that replicated O'Neill's game used a computer environment to generate random choices for subjects (Shachat 2002). The mixed strategy device in this experiment was a computer display of a card shoe, and subjects had 100 cards with which they could fill the shoe with any

TABLE 7.5 OCHS' GAME (1995)

		Player 2	
		Left	Right
Player 1	Up	X, 0	0, 1
	Down	0, 1	1, 0

distribution. Hence, if a player wanted to play a mixed strategy of 60/40, she could select cards in that distribution. Once the cards were placed in the shoe, the computer randomly shuffled them and selected the first one as the strategy that would be played. The opponent only saw the selected card and not the distribution of cards. These experiments, which corrected some of technical flaws in the O'Neill experiment, produced results similar to his results, with the outcome of playing the joker being close to 40 percent.

2. OCHS' EXPERIMENT

While subjects in O'Neill's (1987) experiment could come close to the predicted outcome, other experimental evidence shows that players are unable to learn to play mixed strategies in situations that call for mixed strategy equilibrium. Ochs (1995) had subjects play three matching pennies games, as depicted in Table 7.5. Changes in X in the (Up, Left) cell correspond to different games. In Game 1, X = 1; in Game 2, X = 9; and in Game 3, X = 4.

Varying the X in this game changes the mixed strategy equilibrium outcome for Player 2. When X = 1, Player 1 should mix between Up and Down and Player 2 should mix between Left and Right. When X = 9 or 4, Player 1's strategy should remain the same but Player 2 should play Right 0.9 times when X = 9 and 0.8 times when X = 4.

In Ochs' experiment, subjects played all three games by filling out (or having the computer fill out) a list with 10 strategies for Up and Left. For instance, if a subject listed five Ups and five Lefts, this would be considered a mixed strategy of playing Up half of the time and Left half of the time. In one round, this list of strategies was paired against another subject's list of strategies. Each paired strategy was considered a single game and a winner was determined by comparing each strategy on the lists. In the next round, the subjects again generated a list of 10 strategies, which were again paired with another list of strategies. Subjects were randomly assigned to be Player 1 or Player 2 and were anonymously paired with another subject each round.

Results showed that under these conditions, subjects did not follow a mixed strategy equilibrium for Games 2 and 3. Predictions for relative frequencies were 0.5 for Game 1, 0.9 for Game 2, and 0.8 for Game 3. Experimental results were 0.5015 for Game 1, 0.6309 for Game 2, and 0.5336 for Game 3.

D. PROBABILISTIC CHOICE MODELS

In the prior experiments, we can conclude that subjects could not calculate the precise mixed strategy equilibrium. It is therefore important to ask: Why can't subjects calculate the mixed strategy equilibrium, or are they calculating it and we just can't see it? One concern in experiments is that subjects are human and therefore prone to mistakes. So most subjects might maximize expected behavior, but a few or one might not. These few errant subjects can then bias the results of an aggregate analysis of the data (i.e., when all the subjects' actions are calculated together when measuring behavior).

Both O'Neill and Ochs compared actual frequencies of outcomes with the predicted outcomes. In these cases, they were considering a Nash equilibrium in a literal sense. Recall that a Nash equilibrium is when two best response functions intersect and provide a precise "pinpoint" prediction. When the data are analyzed, the pinpoint predictions often do not match experimental outcomes. But is this discrepancy a result of a few errant subjects who, for whatever reason, do not maximize expected behavior? Some researchers have sought to relax the rigid assumption of the Nash equilibrium when analyzing data to allow for behavior that fundamentally adheres to our definition of expected utility theory but includes some slight errors or a stochastic element attached to choice behavior. This strand of research examines probabilistic choice models that allow for some errors or randomness in the selection of strategies when analyzing experimental data.

In physics, this idea of introducing randomness into models is often referred to as "the uncertainty principle," which argues that certain parameters in a model (such as the precise behavior of subjects) will never be known, so some randomness in the form of a probability distribution predicts both the unknown parameter and the model more precisely (Heisenberg 1927). In quantum mechanics it is a known fact that the precise location of a particle cannot be pinpointed, so there will always be uncertainty about its location. (If you know its position then you do not know its velocity, and vice versa.) However, if it is assumed that particles behave like waves, their location in their quantum state can be specified by a probability distribution, which is a combination of position and velocity. The uncertainty principal, which allows for randomness in the model, is able to provide an accurate prediction of the behavior of particles in experiments.[4] Hawking (1988, 62) notes:

> Quantum mechanics therefore introduces an unavoidable element of unpredictability or randomness into science....[However], it has been an outstandingly successful theory and underlies nearly all of modern science and technology. It governs the behavior of transistors and integrated circuits, which are the essential components of electronic devices such as televisions and computers, and is also the basis of modern chemistry and biology.

McKelvey and Palfrey (1995) use this same notion to model the best response function as a stochastic function, meaning that there is a random component to the payoffs associated with a subject's choice.[5] Just like the best response function discussed earlier, we can think of the intersection of these stochastic (quantal) response functions for each player as an equilibrium, called the quantal response equilibrium (QRE). Palfrey (2006, 928) comments:

> [QRE] makes much different predictions...compared to Nash equilibrium. The QRE model assumes the players are strategic, and are aware other players are also strategic, but players are not perfect maximizers. Rather than always choosing optimally, players choose better strategies more often than worse strategies, but there is a stochastic component to their choices, so inferior strategies are sometimes selected.

In this sense, a QRE is a statistical version of a Nash equilibrium. McKelvey and Palfrey emphasize "that this alternative approach does not abandon the notion of equilibrium, but instead replaces the perfectly rational expectation equilibrium embodied in Nash equilibrium with an imperfect, or noisy, rational expectation equilibrium" (1995, 7).

If we apply a QRE to the Ochs experiments, then we see behavior that more closely measures the behavior predicted by the mixed strategy equilibrium. Recall that the equilibrium predictions for Games 1, 2, and 3 were 0.5, 0.8, and 0.9, respectively, and the actual frequencies were 0.5015, 0.6309, and 0.5336, also respectively. A QRE analysis offers frequencies of 0.746 for Game 2 and 0.669 for Game 3 (McKelvey and Palfrey 1995).

McKelvey and Palfrey (1995) also conducted an analysis of the O'Neill experiment. Recall that the MSE for that game was for the Column and Row players to play the mix (0.2, 0.2, 0.2, 0.4). A QRE found a mix for the Row player to be (0.213, 0.213, 0.213, 0.36) and a mix for the Column player to be (0.191, 0.191, 0.191, 0.426), very close to the actual results. By adding a stochastic element to the notion of a Nash equilibrium, probabilistic choice models such as the QRE model better predict behavior in these types of experiments.

E. TESTING MIXED STRATEGIES USING OBSERVATIONAL DATA

1. SOCCER PLAYERS AND MIXED STRATEGIES

Palacios-Huerta (2003) examined whether the play of soccer (football) players conformed to a mixed strategy equilibrium when taking a penalty shot. In this situation, the kicker can kick the ball to the left, middle, and right of the goal. In an optimal mixed strategy observed over time, penalty kicks should be shot to each of these areas one-third of the time. Palacios-Huerta collected data on 1,417 penalty kicks over five years of games in Spain, Italy, England, and other countries. His data included, among other things, the kicker's leg strength (left or right) and the direction and outcome of the shot (left or right and goal or no goal). He found that soccer players could indeed generate random sequences and "neither switched strategies too often nor too little" (409). He also pointed out that the sport's superstars, such as France's Zinédine Zidane and Italy's Gianluigi Buffon, are remarkable game theorists, since they are able to find a balance among the various strategies, coming close to 33 percent perfection, or achieving equilibrium behavior.[6]

2. TENNIS PLAYERS AND MIXED STRATEGIES

Walker and Wooders (2001) examined top tennis players who have essentially the same strategic set as players facing a penalty kick in a soccer game. The server must decide whether to serve the ball to the right or left court, while the defender must anticipate whether to defend the serve from the left or right court. Walker and Wooders obtained videotapes of 10 Grand Slam tournaments and the year-end Master's tournaments and analyzed 160 first serves[7] and then compared their results to the results of O'Neill's (1987) experiment. They found strong evidence that tennis professionals generally followed a mixed strategy equilibrium and noted (2001, 1535):

> There is a spectrum of experience and expertise, with novices (such as typically experimental discerns) at the one extreme and world-class tennis players at the other. The theory applied well (but not perfectly) at the "expert" end of the spectrum, in spite of its failure at the 'novice' end.

One lesson to learn from these experiments and observational studies is that the "professional" RPS competitors discussed at the beginning of the chapter have unique skills that mirror skills in other professional sports where randomization is a key to strategy selection.

F. SUMMARY

In this chapter we learned how to calculate mixed strategies, where these strategies are a probability distribution over choices. When both players calculate their optimal mixed strategy, a mixed strategy equilibrium occurs, since the calculated mixed strategies are a best response to each other and a player should be indifferent between playing each strategy. In theory, the use of mixed strategies assumes that players use a randomizing device to make play optimal. However, experiments have shown that real players often have a difficult time playing mixed strategies in their heads. Consequently, many experiments allow subjects to use a random device when implementing mixed strategies in the laboratory. Both field and experimental data have shown that mixed strategies resulting in equilibrium outcomes do occur "weakly" in the laboratory and in the field.

Thus far, we have only considered simultaneous move games in which players are unaware of how their opponent has selected before making their own selection. In the next chapter, we will begin to examine dynamic games of complete information, in which a player knows what strategy another player has selected before he or she must decide on a strategy choice.

NOTES

1. The cartoon *The Regular Show* features two friends, a bluejay named Mordecai and a raccoon named Rigby, who are always trying to get out of work. The friends always use RPS to settle disputes, but Mordecai always plays rock and Rigby always plays scissors. Mordecai has figured this out, so he always suggests RPS in a dispute, and Rigby, who hasn't figured it out, always agrees and always loses.
2. Experiments have shown that when both players play RPS blindfolded, they tend to make more random decisions than when only one player is blindfolded. The difference between the two cases is statistically significant. Cook et al. (2011) argue that this is evidence that players in some decision-making environments tend to mimic their opponents' actions. This mimicking behavior is referred to as automatic imitation.
3. You should verify that the expected value of the game for both players is 0.66.
4. Einstein, although credited with founding the field of quantum mechanics, never fully accepted it as a valid science because of its randomness. Recall his famous objection: "God does not play dice" (Hawking 1988, 62).
5. There are many different types of models of this nature. A few are the cognitive hierarchy model, the agent quantal response equilibrium model, the cursed equilibrium model, and the noisy Nash model. See McKelvey and Palfrey (1995; 1998) and Goeree, Holt, and Palfrey (2008) for examples.
6. Using the same soccer data, Woolders (2010) found evidence to the contrary. His results found that players tended to overplay and underplay in different segments of the game, so that strategies were not random.
7. They also analyzed second serves, but the sample was generally small.

CHAPTER 8

EXTENSIVE FORM GAMES AND BACKWARD INDUCTION

A. THE TWENTY-ONE COIN GAME

Consider the basic 21 coin game, in which two players each have a pile of 21 coins in front of them. Taking turns, each player can remove one, two, or three coins at a time. The player who removes the last coin is declared the winner. This game has a unique solution, and the player who selects first should win. If you were the first player, how would you play?

To solve this game, let us consider its end. On the penultimate move, you want your opponent to have four coins, so that he will have at least one coin left after playing. How can you ensure that this occurs? We can reason backwards from the end of the game to infer that in order to leave him with four coins on the penultimate move, you have to leave him with eight coins on the turn before, 12 coins on the turn before that, 16 coins before that, and finally 20 coins before that. Therefore, starting with 21, the first player should remove one coin and then proceed to remove a number of coins equal to four minus the number the other player removes during his turn. So if in the first move you take one and your opponent takes two, you should take (4 − 2) or two coins on your next move, and so on. This strategy will ensure that you will win the game.

This simple game is relevant to this chapter. In the next section, I will define an extensive form game and give some examples of how to construct one, and in the following section, I will show how to solve these games using the method of backward induction. Next, I will illustrate a first and second mover's advantage in these games. Last, I will present several experiments in which subjects attempt to solve games like the 21 coin example given here.

B. DEFINING AN EXTENSIVE FORM GAME

1. FOLLOW-THE-LEADER GAME REDUX

Thus far we have solved games using their normal form, but more interesting and complex games can be constructed using the extensive form, or by building a game tree. A game tree consists of a series of nodes linked in a sequence. Each node has a number of branches that lead to other nodes and branches that eventually end at a terminal node. In this chapter, we will consider complete and perfect information extensive form games in which players know where they are located in the game tree (i.e., they know the node at which they are located). Chapter 10 will examine imperfect information games in which a player may not know where she is located in the game tree prior to making her choice. Complete information extensive form games are useful in modeling sequential decision-making such as bargaining.[1]

To show how the mechanics of an extensive form game tree work, consider the game displayed in Figure 8.1, in which three players must decide which direction to select. Player 1 can make his decision first, then Player 2, and finally Player 3. All players receive a payoff of 1 if they all select the same direction; otherwise, their payoff is 0.

Notice that all three players have a choice of the same two strategies of Left or Right. Player 1's choices are denoted by L1 and R1, Player 2's choices are denoted by L2 or R2, and Player 3's choices are denoted by L3 or R3.

The sequential nature of the game means that all three players have different decisions to make depending on the node at which they are located. First, notice that some nodes are contained in the tree (decision or choice nodes) while other nodes lie at its ends (terminal nodes). At these nodes, no further decisions are made and the payoffs for the three players are listed, with Player 1's payoff at the top, Player 2's in the middle, and Player 3's at the bottom. Notice that at each decision node, a player has a choice of two decisions to make—left or right—but each player confronts a different decision depending on the particular node at which he or she is located.

Since Player 1 moves first, she only has one node and a choice of left or right. Player 2 moves next and faces two decision nodes: one when Player 1 selects R1 and the other when Player 1 selects L1. Therefore, if Player 1 plays L1, then Player 2 can play L2 or R2, and if Player 1 plays R1, then Player 2 can play L2 or R2. Finally, Player 3 faces four decision nodes, or eight possible decisions, depending on what Player 1 and Player 2 select.

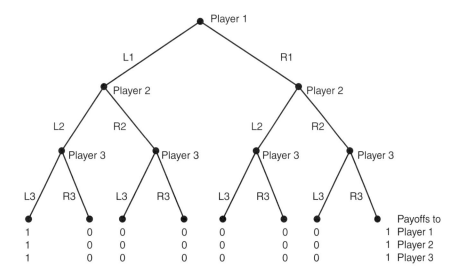

FIGURE 8.1 Follow-the-leader game

In Figure 8.1, I have labeled Player 2 at each of his two decision nodes and Player 3 at each of her four decision nodes. It is important to understand that each node in a row represents a single player's decisions. In subsequent examples, to remove clutter, I will not label players across nodes; instead, I will label them on the right side of the nodes, and it should be assumed that the nodes in a row represent all of a single player's potential decisions at that point in the tree.

The solution to this game is trivial, since it is a complete and perfect information game and we assume our players are rational. Consequently, this is a follow-the-leader game: if Player 1 selects L1 (or R1), then Player 2 and Player 3 should select L2 and L3 (or R2 and R3), respectively. In this case, the players win and each receives a payoff of 1. Any errant choice by Player 2 or Player 3 would result in a loss for all players, but complete and perfect information and rationality rule out these types of mishaps. In Chapter 10, we will revisit this game with players who must make decisions under conditions of incomplete information. As we will see, the assumption of incomplete information changes this game from a follow-the-leader game to a guessing-matching game in which players have to guess how another player has moved. That is, a player does not know the node at which he is located when he makes a decision. In this chapter and the next, we assume that players have complete information about the move made by another player ahead of them in the game tree (i.e., they know the node at which they are located when they make a decision).

2. FORMAL DEFINITION OF EXTENSIVE FORM GAME

An extensive form game with complete and perfect information requires:

1. A set of players.
2. A set of sequences of nodes called decision (or choice) nodes, with the ending nodes referred to as terminal nodes.
   a. Each terminal node can only have one outcome for each player.
   b. The extensive form game cannot have more than one sequence of paths that leads to an individual terminal node.
   c. The extensive form game cannot have a sequence of choice nodes that lead to a node prior to their origin.
   d. A single node (one that is not attached to another node) is called a singleton. For all complete information extensive form games, nodes are considered singletons so that players can know where they are in the game.
3. Perfect recall is assumed, meaning that players cannot forget how they have moved previously.[2]
4. For each sequence of choices, there is a set of outcomes for which players have a utility function defined over the outcomes.

These "rules" provide a standard way to represent extensive form games so that they can be analyzed and solved in the same way that we solve normal form games. We will discuss some of these rules in Chapter 10. To understand how an extensive form game is created and how payoffs are assigned to actions, I will now provide two examples. Simply focus on the structure of the game in these examples. We will discuss how to solve these types of games in section C.

3. *TWILIGHT* EXAMPLE

The first example I call the *"Twilight* example," in homage to the popular young adult book and movie. In the *Twilight* saga, a teenage girl and a vampire have fallen in love with each other. Realizing that vampires are immortal, the girl wants the vampire to suck her blood so that she will also become a vampire and the two can spend their immortal lives together. However, the vampire does not want to turn the girl into a vampire, because he feels that vampires (himself included) are monsters, and he does not want the girl to become a monster. However, he truly loves the girl and wants to be with her for as long as he can. I will first model this situation in an extensive form game by specifying each move and strategy set, and I will then examine payoffs or utilities for each action.

The vampire has the first move and has two choices: to bite the girl (B) or to not bite the girl (NB), as illustrated in Figure 8.2.

If NB is selected, the game ends. If the vampire decides to suck the girl's blood (selecting B), the girl has the second move, illustrated in Figure 8.3, of whether to resist being bitten (R) or to not resist being bitten (NR). If R is selected the game ends.

If the girl decides to not resist being bitten (selecting NR), the vampire has the final move of whether to marry the girl (M) or to not marry her (DM), as shown in Figure 8.4.

Now that we have specified the players' strategies and sequence of moves, we can attach payoffs to the game. To calculate appropriate payoffs, we define ordinal utilities for each outcome a player has and make judgments about the consequences of those outcomes relative to the other player. Consider the vampire's first move. If he does not bite the girl, she will eventually die of natural or other causes and he will have to live the rest of his immortal life without her, but at least he will be alive. We can assign a payoff of 5 to the vampire, since he will at least have memories of the girl after she dies. The girl will have spent her natural life with the man she loves, but it is unclear how their relationship will continue as she grows older, since the vampire will remain a teen and he might not be able to love an elderly woman. We can assign a lower payoff of 3 to the teenage girl, since she will die before the vampire and she is uncertain about how long

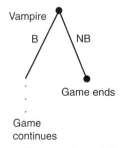

FIGURE 8.2 First move in *Twilight* example

they will have a relationship. On the second move, if the teenage girl chickens out and resists, then the vampire will be rejected and the couple will break up. Thus, each player will gain a negative payoff of −1, since they prefer to be together rather than apart. In the vampire's last's move, he has the option of either marrying the girl or not marrying the girl. If he decides to not marry the girl because her behavior has changed as a result of being turned into a vampire, he can simply move on to find another mate, so his payoff will be 0. However, if this occurs, the teenage girl has been turned into a vampire, is now a monster, and does not get to be with the man she loves. This is the worst outcome for the girl and can be assigned a payoff of −5. If the vampire decides to marry her, this decision offers the best outcome for both of them: they will live out their immortal lives together and each will get a payoff of 10. Figure 8.5 contains the extensive form with payoffs, with the top payoff being the vampire's payoff and the bottom payoff being the teenage girl's payoff.

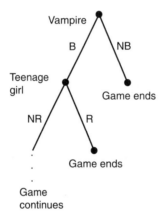

FIGURE 8.3 Second move in *Twilight* example

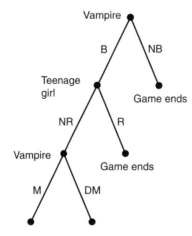

FIGURE 8.4 Final move in *Twilight* example

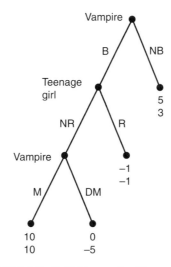

FIGURE 8.5 *Twilight* example with payoffs

4. THREE STOOGES GAME

Consider a three-person game with the Three Stooges Larry, Moe, and Curly. This comedic trio was popular during the era between the 1930s and the 1950s. Their routine usually centered on physical slapstick humor, with one always ending up hurt. In this game, Larry can either slap Moe (SM) or not slap Moe (DM), Moe can either slap Curly (SC) or not slap Curly (DC), and Curly can either slap Larry (SL) or not slap Larry (DL). In this game, each Stooge gains utility from watching another Stooge be slapped, with utility increasing as the number of slaps increases. However, none of the Stooges want to be slapped, and they gain negative utility if they are. Also, since Larry is the leader of the group, the other Stooges gain greater utility from watching him get slapped, and Larry is humiliated more by being slapped, so he gets greater negative utility than the others. If a Stooge is not slapped, the utility is zero. When slapping occurs, Moe and Curly get a utility of 1 from watching each other get slapped, but they get a utility of 2 from watching Larry get slapped. If Moe or Curly gets slapped, they get a utility of −1, and if Larry gets slapped, he gets a utility of −2. Hence, if Larry slaps Moe, Moe slaps Curly, and Curly slaps Larry, then the payoffs are as follows: Larry gets 0, since he gets 1 from watching Moe get slapped, 1 from watching Curly get slapped, and −2 from getting slapped himself (1 + 1 − 2 = 0). Moe gets −1 from getting slapped, 1 from watching Curly get slapped, and 2 from watching Larry get slapped, for a payoff of 2. Curly receives the same payoff as Moe. If no one gets slapped, each Stooge gets a payoff of 0. The best payoff for both Moe and Curly is not getting slapped and watching the others get slapped (producing a payoff of 0 + 1 + 2, or 3). Larry's highest possible utility is only 2, since he gets a utility of 1 from watching Moe and Curly get slapped. I will leave it to the reader to calculate the other payoffs. This game is represented in Figure 8.6, with Larry going first, Moe going second, and Curly going third. The payoffs listed under each terminal node represent the order in which the players move, with Larry first, Moe second, and Curly last.

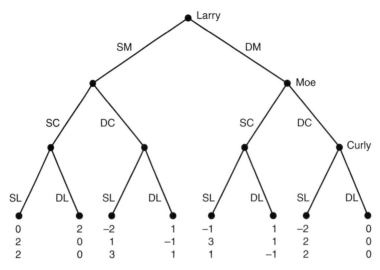

FIGURE 8.6 Three Stooges game

C. BACKWARD INDUCTION

One method to solve extensive form games with complete information is backward induction (sometimes called a rollback equilibrium).[3] Since all players have complete and perfect information about each move's strategies and associated payoffs, as well as all prior moves, each player can deduce what another player will do at all decision nodes. In a three-player game, Player 1 can examine the payoff structure, reason what Player 3 will do on her last move, and then deduce what Player 2 will do. That is, Player 1 can deduce what Player 2 will do at each decision node and what Player 3 will do at her terminal node. To solve the extensive form with complete and perfect information, we start at the terminal node for Player 3, examine her payoffs at each node, and decide what action she will take at each node. Then, given her actions, we can determine what Player 2 will do, and finally, Player 1 can pick the strategy that leads to her optimal outcome. To see this progression, consider the game depicted in Figure 8.7. In this game, Player 1 first selects L1 or R1, then Player 2 selects L2 or R2, and finally Player 3 selects L3 or R3. To solve this game using backward induction, let us first consider Player 3. Recall that her payoffs are the final entry in the list of the three payoffs.

Let us examine the first set of choices, in which Player 1 selects L1, Player 2 selects L2, and Player 3 must choose between L3 and R3. In this case, she has a choice between L3, which yields a payoff of 50, or R3, which yields a payoff of 40. Player 3 will obviously select L3 over R3, since this choice gives her a payoff of 50 as opposed to 40. We can thus eliminate R3 from this node. If we examine the next set of choices, in which Player 1 selects L1, Player 2 selects R2, and Player 3 is choosing between L3 and R3, we can see that L3 will yield 30 and R3 will yield 20, so Player 3 will select L3 and we can eliminate R3 from this node also. In the case in which Player 1 selects R1, Player 2 selects R2, and Player 3 has a choice between L3, which offers a payoff of 60, and R3, which offers a payoff of 80, it is

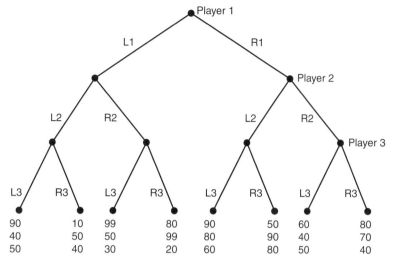

FIGURE 8.7 Solving for backward induction

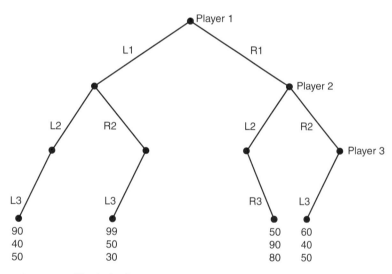

FIGURE 8.8 First induction

clear that she will select R3, so we can eliminate L3 from this node. Finally, when Player 1 selects R1 and Player 2 selects R2, Player 3 faces a choice between L3, which yields 50, and R3, which yields 40, so she will select L3 and we can eliminate R3 from this node. We can now eliminate all of the strategies that Player 3 will not select, producing the pruned tree depicted in Figure 8.8.

Now we turn to Player 2. If Player 1 selects L1, then Player 2 knows that if he selects L2, Player 3 will select L3, and if he selects R2, Player 3 will select L3. Selecting R2 yields Player 2 a payoff of 50, as opposed to the payoff of 40 that will result from selecting L2, so if Player 1 selects L1, then Player 2 will select R2. Now, if Player 1 selects R1, Player 2 knows that Player

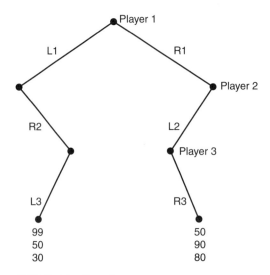

FIGURE 8.9 Pruned tree

3 will select R3 if he plays L2 and L3 if he plays R2. Since R3 yields the highest payoff (90 as opposed to the 50 that will result from L3), Player 2 will play L2. We can now eliminate the strategies that Player 2 will not play. Figure 8.9 depicts the game tree that Player 1 actually confronts.

Player 1 knows that if she selects L1, Player 2 will select R2 and Player 3 will select L3, yielding a payoff of 99, and she knows that if she selects R1, Player 2 will select L2 and Player 3 will select R3, yielding a payoff of 50. Consequently, the equilibrium path for this game is for Player 1 to select L1, Player 2 to select R2, and Player 3 to select L3, resulting in payoffs of (99, 50, 30). The equilibrium path specifies the behavior of players in equilibrium.

Now let us solve the two games we first presented in this chapter. What are the solutions? In the *Twilight* example, the vampire will bite the girl, she will not resist, and they will get married and live happily ever after. In the Stooges game, Larry, Moe, and Curly will all slap each other (is this a game of coordination or conflict?).

D. THE IMPORTANCE OF THE ORDER IN WHICH PLAYERS MOVE

1. FIRST MOVER'S ADVANTAGE AND THE CHICKEN GAME

In extensive form games, the sequencing of moves can have an impact on outcomes. That is, in some games, the order of who goes first, second, third, and so on matters. Consider the chicken game discussed in Chapter 6 and reproduced in Table 8.1.

Before I get to the point of this section, I want to step back a moment to show how to convert normal form games into extensive form games and vice versa. This issue will become important in the next chapter when I discuss subgame perfect equilibrium. But for now, let us take the normal form game of chicken depicted in Table 8.1 and convert it to an extensive

TABLE 8.1 CHICKEN GAME

		Teen 2	
		Swerve	Don't swerve
Teen 1	Swerve	1, 1	0, 2
	Don't swerve	2, 0	−1, −1

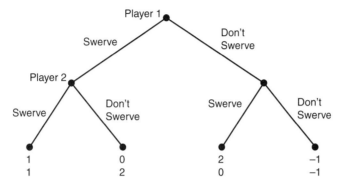

FIGURE 8.10 Extensive form of the chicken game

form game. First, realize that we are concerned with games of perfect and complete information, so we are converting a simultaneous move game into a sequential game in which we know all of players' strategies, payoffs, and moves. As such, we have to make a decision about who moves first and who moves second. We will see that the order in which players move can often dictate the outcome of the game. For illustrative purposes, let us assume that Player 1 moves first. In this case, Player 1 has two strategies: Swerve and Don't Swerve. Player 2 also has two strategies of Swerve and Don't Swerve, but since this is now a sequential game, those strategies are contingent on the strategy of Player 1. As a result, Player 2 has four decision nodes: if Player 1 selects Swerve, Player 2 can select Swerve or Don't Swerve, and if Player 1 selects Don't Swerve, then Player 2 can select Swerve or Don't Swerve. This extensive form is represented in Figure 8.10

Note that the normal form game has two equilibria: the first is when Player 1 selects Don't Swerve and Player 2 selects Swerve, and the second is when Player 2 selects Don't Swerve and Player 1 selects Swerve. The extensive form, however, has only one equilibrium, in which Player 2 does the opposite of what Player 1 selects. Conducting a backward induction on this game reveals that Player 1 has the advantage of the first move, since his choice of Don't Swerve demands that Player 2 select Swerve.

2. FIRST MOVER'S ADVANTAGE AND A COLLECTIVE GOOD GAME

Let us revisit the collective goods game discussed in Chapter 6. A revised version of this game can be represented as a three-player sequential move game that concerns whether or not a person should contribute to a collective good. Assume that at least two players must contribute money to produce some collective good. Therefore, each player has a decision of whether to contribute money to the collective good (C) or to not contribute money to the

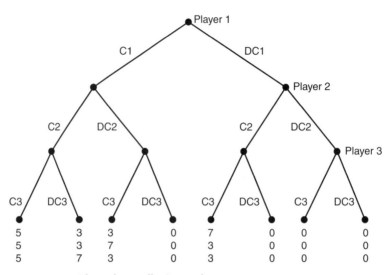

FIGURE 8.11 Three-player collective goods game

collective good (DC). If all three players contribute to the collective good, then a collective good is established at the optimal level of 5. However, if only two players contribute to the collective good, then the good is established at a level of 3, below the optimal level. In this case, the player who does not contribute is able to free-ride off the other players. Because the collective good is established at a level of 3 but the player who does not contribute gets to enjoy the collective good for free without investing any of her resources, her utility is 7. Also note that when two or more players do not contribute to the collective good, then all players receive a payoff of 0, since the collective good cannot be established. The extensive form of this game is presented in Figure 8.11.

In this game, the first mover (Player 1) has an advantage, since she can simply choose DC1 and force Players 2 and 3 to select C for an equilibrium strategy (DC1, C2, C3) and payoffs of (7, 3, 3). Hence, being a first mover allows Player 1 to achieve her highest payoff of 7 while free-riding off of the other two players.

3. SECOND MOVER'S ADVANTAGE AND RPS GAME

To consider a second mover's advantage, I present a complete information version of the rock-paper-scissor game discussed in the last chapter. In this game, two players must play Rock, Paper, or Scissors, where Rock beats Scissors, Paper beats Rock, and Scissors beats Paper. The associated payoffs from this game are depicted in Figure 8.12.

When we do a backward induction on this game, notice that when Player 2 is confronted with Scissors, she will select Rock; when confronted with Paper, she will select Scissors; and when confronted with Rock, she will select Paper. Player 1's strategy is always losing. Moving first tips Player 1's hand, demonstrating why mixed strategies are important in these types of games.

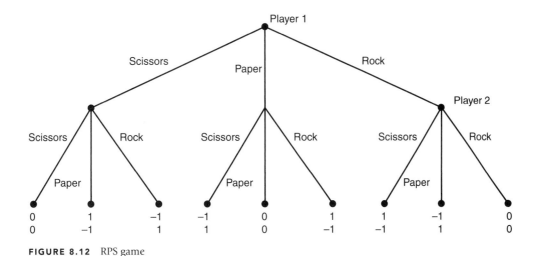

FIGURE 8.12   RPS game

E. BACKWARD INDUCTION AND THE NEED FOR REFINEMENT

Backward induction always identifies the Nash equilibrium of the game but does not identify *all* of the Nash equilibria if multiple exist. In the chicken game outlined in Table 8.1, the normal form game identified two equilibria, but the backward induction solution only identified one (Don't Swerve, Swerve). What happened to the other equilibrium? It is still there, but it does not make much sense in the context of the extensive form. The other equilibrium is when Player 1 selects Swerve, giving Player 2 the chance to select Don't Swerve. But this action reduces Player 1's payoff from 2 to 0, so it would not be a rational choice. As a result, this equilibrium is suspect in the extensive form game.

Many extensive form games contain equilibria that do not make sense in terms of the strategies that a Nash equilibrium demands that a player play. For example, consider the game depicted in Figure 8.13, in which Player 1 has the option of selecting L1 or D1.[4] If Player 1 selects L1, then the game ends and Player 2 does not get to move. The only way that Player 2 gets to move is if Player 1 selects D1. When Player 1 selects D1, then Player 2 has a choice between L2 and R2. If Player 2 is allowed to move, he will select L2 and get a payoff of 1, as opposed to selecting R2 and getting a payoff of 0. Player 1 knows that if she selects D1 she will get a payoff of 3, and that if she selects L1 she will get a payoff of 2. The backward induction solution reveals that (D1, L2) with a payoff of (3, 1) is the Nash equilibrium to this game.

This game has another Nash equilibrium, which is revealed by examining the normal form of the game, as shown in Table 8.2. In this game, (L1, R2) with a payoff of (2, 2) is also a Nash equilibrium. However, notice that this equilibrium requires Player 2 to select R2 when Player 1 selects L1. But recall that if L1 is selected, then Player 2 does not get to move. In this case, there is no way to specify Player 2's behavior, since he has no move when L1 is selected. A Nash equilibrium only specifies behavior when a player has an action to take. The behavior of players whose actions are not part of an equilibrium is referred to as behavior off the equilibrium path. That is, in Figure 8.13, the equilibrium path for the equilibrium (L1, R2) is represented

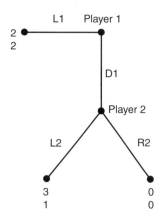

FIGURE 8.13 Equilibrium that doesn't make sense

TABLE 8.2 NORMAL FORM OF THE EXTENSIVE FORM GAME DEPICTED IN FIGURE 8.12

		Player 2	
		L2	R2
Player 1	L1	2, 2	2, 2
	D1	3, 1	0, 0

by the line from Player 1's decision node to the terminal node with payoffs (2, 2), and this line does not connect with Player 2's decision node. A Nash equilibrium is only concerned with moves that are on the equilibrium path. Since the equilibrium path does not go through Player 2's decision node, he cannot strictly prefer L2 over R2. The equilibrium (L1, R2) specifies that Player 2 will select R2, but if his decision node is reached, he will select L2. So this equilibrium makes no sense, because it does not consider behavior off the equilibrium path.

Also notice that Player 2 would prefer that Player 1 choose L1, but the only way that Player 2 can get Player 1 to choose L1 is to threaten Player 1 that if she selects D1, he will play R2. But this is an empty threat, and if Player 1 plays D1, then Player 2 will be forced "rationally" to play L2. Hence, this Nash equilibrium is only supported by an empty threat and is an unreasonable equilibrium that should be eliminated.

Backward induction eliminates unfeasible equilibria because, unlike the Nash equilibrium, it considers moves that occur on and off the equilibrium path. As a result, this method will not find equilibria that are supported by irrational behavior. That is, backward induction does not identify all of a game's equilibria, but only selects the "right" equilibrium. In the next chapter, I will present a subgame perfect equilibrium, which is a refinement of backward induction.

F. EXPERIMENTS ON BACKWARD INDUCTION REASONING

1. RACE GAME

Backward induction is a fundamental procedure of game theory and so it is natural to question people's ability to perform this procedure in their minds. Recall our 21 coin game.

Solving this simple game requires the cognitive ability to use backward inductive reasoning. For behavioral game theorists, the question is whether subjects in a laboratory setting can discern solutions to these types of games. Gneezy, Rustichini, and Vostroknutov (2007) conducted an experiment on a game called "the race game." They considered two games, one called G(15, 3) and another called G(17, 4). The solution to these games is like our 21 coin solution. In the G(15, 3) game, two subjects start out by selecting numbers in intervals of three, and the first one who arrives at 15 wins. The solution is for a player to get to 11 first, so that the opponent can only play 12, 13, or 14 in the next round, and the player will then play 3, 2, or 1 and win. The G(17, 4) game (in which subjects try to get to the number 17 by selecting numbers in intervals of four) is solved in the same way, but a player is trying to get to 12 first. In this case, the opponent can only play 13, 14, 15, or 16, and the player can then play 4, 3, 2, or 1 and win.

In Gneezy, Rustichini, and Vostroknutov's (2007) experiment, subjects were randomly and anonymously matched with other subjects in a session. Each subject played G(15, 3) for 20 rounds, with the winner getting 20 dollars and the loser getting nothing. The G(17, 4) game was played 10 times in some of the sessions.

The experiment found that in the G(15, 3) treatment, subjects did learn the optimal solution by the twentieth round, with the mean deviation[5] from the optimal solution being 0.535 in the first 10 rounds and only 0.124 in the second 10 rounds. That is, by round 20, the subjects had learned the backward induction solution to the game. Similar findings were found for the G(17, 4) treatment. What these experiments showed is that subjects are able to learn how to do backward induction, but it takes time for them to learn the optimal solution.

Dufwenberg, Sundaram, and Butler (2008) conducted experiments on a different 21 coin game than was illustrated in the introduction to this chapter. These researchers were interested in the question of whether the use of a simpler game prior to playing the 21 coin game would help subjects solve the more complex 21 coin game. In essence, they gave subjects a short tutorial on how to solve games of this type and then let subjects play the game. Using Gneezy, Rustichini, and Vostroknutov's (2007) notation and game format, they used two games called G(6, 2) and G(21, 2). In G(6, 2), the second player has the advantage and can win by selecting 3, and in G(21, 2), the first player can win by selecting 1 and having the opponent reach 18 first. Dufwenberg, Sundaram, and Butler conducted two treatments: in the first treatment, subjects played G(6, 2) for five rounds and then played G(21, 2) for five rounds, and in the second treatment, subjects played G(21, 2) for five rounds and then G(6, 2) for five rounds. The idea was that playing the simpler G(6, 2) first would give subjects insight into how the solution was calculated, allowing them to perform better in the harder G(21, 2) game.

Subjects were students in one of the author's microeconomics courses. Dufwenberg, Sundaram, and Butler note (2008, 7):

> Since subjects were in class, we had no reason to make sure each was compensated for their time. We used a pay-a-random-subset-of-subjects approach . . . : two subjects from each treatment were selected at random (one for G21; one for G6) and paid $5 for each game won.

In this experiment subjects were not anonymously matched with other subjects, but they were unable to communicate with each other. The researchers found evidence that playing $G(6, 2)$ prior to $G(21, 2)$ allowed subjects to better solve $G(21, 2)$. When $G(6, 2)$ was played first, then 37 percent of $G(21, 2)$ were perfect games, but when $G(6, 2)$ was played second, only 21 percent of $G(21, 2)$ were perfect games. As in the Gneezy, Rustichini, and Vostroknutov (2007) experiment, these findings show that subjects are able to do backward induction reasoning, but they require a learning period. Dufwenberg, Sundaram, and Butler (2008, 19) comment, "While humans have a language instinct with which to acquire proficiency in spoken language, strategic thinking, like written language, has to be learned the hard way."

2. RACE GAME AND CHESS PLAYERS

The two previous experiments concerned "normal subjects." Levitt, List, and Sadoff (2009) wondered how chess experts would perform in solving these types of simple backward induction games. Chess is a game in which players must deduce how a single move will affect game play several moves down the line. Hence, the cognitive processes of chess players, especially masters of the game, must be attuned to backward induction. Palacios-Huerta and Volij (2009, 1624) comment:

> Backward induction reasoning is second nature to expert chess players. They devote a large part of their life to finding optimal strategies for innumerable chess positions using this reasoning. Further, it is common knowledge among them that they are all highly familiar with backward induction reasoning.

Levitt, List, and Sadoff (2009) conducted experiments at two international open chess tournaments in 2008, the Chicago Open and the World Open. They rented a room at the convention center that was hosting the tournament and conducted two "race to 100" games with Grandmasters, International Masters, and lesser chess players. Using the notation and game format of Gneezy, Rustichini, and Vostroknutov (2007), they conducted two games called $G(100, 9)$ and $G(100, 10)$. These games were harder to solve than games in prior experiments, since they required players to reason back 10 moves.

In $G(100, 9)$, the second mover has an advantage, since the first mover cannot select 10, which is the optimal move, since the solution involves selecting 10, 20, 30, and so on, up to 100. In $G(100, 10)$, the first mover has an advantage, since the solution is 1, 12, 23, 34, 45, 56, 67, 78, 89, 100.

Levitt, List, and Sadoff (2009) matched two chess players anonymously so that participants did not know each other's identity. The winner was paid 10 dollars and the loser received nothing. Results showed that in the (100, 9) game, 57.3 percent of players solved the game on the first move (i.e., at 10), and 66 percent solved the game by the second move. In contrast, in the (100, 10) game, only 12.6 percent of subjects solved the game on the first move and only 22 percent solved the game by the second move. Why the discrepancy? The authors note (2009, 13):

> We find it striking that even among a subject pool that has extensive experience with backward reduction, the seemingly minor change of shifting the "key numbers" from numbers ending with zero leads to a sharp reduction in success in solving the problem. The result is

consistent with the power of subtle changes reported in many psychology experiments as well as Binmore et al. (2002, p. 87), who find that backward induction behavior of players unfamiliar with the game is quite sensitive to minute changes in the game, and also with the findings of Adriaan de Groot (1965) regarding the difficulty chess players have in generalizing their skills in unfamiliar settings, even within relatively narrow contexts.

What the experiments in this section show is that subjects in controlled experiments do have the ability to do backward induction reasoning, but this skill is not natural and requires learning. In simple games, we would expect subjects to perform well, but in more complex games, we would expect subjects to perform poorly until they had gained considerable knowledge about how solutions are derived.

G. SUMMARY

In this chapter, I showed how to construct extensive form games by specifying strategy sets at each node and payoffs available to players for each action. I also explained that the equilibrium to these games can be solved by using backward induction. Experiments have shown that both normal subjects and experts can perform the backward induction calculations, but there is a learning curve for them to find the optimal solution. In the next chapter, we examine the first refinement of a Nash equilibrium, subgame perfect equilibrium.

NOTES

1. Von Neumann (1928) was the first to adopt this structure in analyzing games
2. I will give an example of imperfect recall in Chapter 10.
3. This method is used in Zermelo (1913) and Kuhn (1953).
4. This example is taken from Selten (1965) and Fudenberg and Tirole (1991).
5. Meaning how close they are to solving the game. Smaller deviations indicate closer proximity to the solution than larger deviations.

CHAPTER 9

SUBGAME PERFECT EQUILIBRIUM

A. CREDIBLE VS. NONCREDIBLE THREATS

The idea of credible versus noncredible threats is an important subject in extensive form games. What do we mean by these terms? Consider the following examples:

1. A family is staying at a Disney resort for a week, and on the first day the kids are behaving horribly. The father tells the children that if they do not stop misbehaving, they are going to go home. Is this a credible threat?
2. A child wants to go outside and play in the rain, but her father refuses to let her go out, because he fears the child will get sick. The child threatens that she will hold her breath until the father relents. Is this a credible threat?
3. You are in a fairly populated market during the day and someone approaches you with no disguise, pulls a gun out, and says, "Give me your money or I will shoot you." Is this a credible threat?
4. It is night and you are walking home with no one around. Suddenly, someone jumps out from behind a bush wearing a mask. You can see a gun pointed at your face and you hear a voice say, "Give me your money or I will shoot you." Is this a credible threat?
5. You are a bank teller and a man enters the bank with what appears to be a bomb attached to his body and threatens to detonate the bomb if you do not give him money. Is this a credible threat?
6. Two nations with nuclear capabilities are disputing territory in international waters and one nation says to the other, "If you do not grant me the deed to this territory, my nation will attack your nation with nuclear weapons." Is this a credible threat?

In these six examples, which ones actually illustrate a credible threat? The first two examples are not credible, since there is no way that a family would go home from a luxury vacation at Disney because the children are misbehaving, and there is also no way that a child could hold her breath long enough to force her father to give in to her demand. Examples 3 and 6 are semi-credible threats, since a gunman might shoot you in a crowded market even with witnesses around, and it is possible that a nation would attack another nation with nuclear weapons over disputed territory in international waters. However, Examples 4 and 5 pose the most serious threats, since a bank robber with a bomb attached to his body might be crazy enough to detonate it, and a robber with a gun, a concealed identity, and no witnesses might shoot you if you refused to give up your money.

In this chapter, we will discuss a refinement to the backward induction Nash equilibrium concept that eliminates equilibria that are sustained by noncredible threats. By incorporating a complete strategy plan (or mapping) for players, subgame perfect equilibrium is able to rule out equilibria that are deemed unreasonable, such as those supported by noncredible threats. In the next sections, I will define a subgame and we will learn how to calculate a subgame perfect equilibrium. We will also discuss some of the limitations to this approach. In the fourth section, I will discuss experimental tests, starting with the centipede game. In the fifth and sixth sections, I will present ultimatum and trust games, respectively, and then discuss neuroeconomics, which attempts to examine subjects' brain functions to measure emotions. Finally, I will discuss the credibility of the experimental findings.

B. SUBGAME PERFECT EQUILIBRIUM

1. SUBGAMES

As the last chapter noted, the Nash equilibrium can often produce equilibria that do not make sense "rationally." Recall that rationality requires that players on the equilibrium path select strategies that maximize their expected payoffs. But Nash equilibrium has nothing to say about behavior when it is off the equilibrium path. One reason for this situation is that Nash equilibrium is a normal form concept that ignores the sequential structure of game play that appears in extensive form games. The subgame perfect equilibrium for sequential games allows us to get rid of "irrational" equilibria by specifying a complete plan that is contingent on an opponent's strategies. These types of strategies are often referred to as behavioral strategies (Kuhn 1953; Selten 1975).

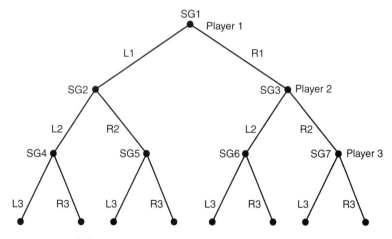

FIGURE 9.1 Subgames

A subgame of an extensive form game starts at a single node and includes all subsequent nodes that compose a game.

Figure 9.1 contains seven subgames ranging from SG1 to SG7: the initial node, the nodes after Player 1's L1 and R1 moves, and the four nodes after Player 2's L2 and R2 moves. A subgame perfect equilibrium involves rational play at every subgame where there is an equilibrium. For example, let us assume that there is one Nash equilibrium in subgame SG4 and another one in subgame SG6. A subgame perfect equilibrium requires players to play rationally in both of these subgames. If players are not behaving rationally in these subgames, then equilibrium is not subgame perfect. We can thus define a subgame perfect equilibrium as a Nash equilibrium that requires all players in the game to play rationally in every subgame.

What is considered irrational play in a subgame? If a player in a subgame is selecting a lower payoff over a higher available payoff, then the player is behaving irrationally and that choice is not subgame perfect.

2. THREAT GAME

To understand this concept further, consider the threat game shown in Figure 9.2 and its normal form expressed in Table 9.1. Notice that the game in Figure 9.2 has two subgames: the whole game and the game starting at Player 2's node.

This game has two Nash equilibria, denoted by the asterisks in the (R1, L2) and (L1, R2) cells. In this game, for Player 2 to achieve his highest utility of 2, Player 1 must play R1, but Player 1 has no incentive to play R1 since that action will only get her a payoff 0. Player 2 could threaten Player 1 that if she does not play R1, he will respond by playing L2 instead of R2, resulting in a payoff of −1 for both of them. If Player 1 thinks that this is a credible threat, then Player 1 might have an incentive to give in and play R1. But is this

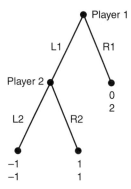

FIGURE 9.2 Threat game

TABLE 9.1 NORMAL FORM OF THREAT GAME

		Player 2	
		L1	R1
Player 1	L1	−1, −1	1, 1*
	R1	0, 2*	0, 2

threat credible? Consider the extensive form game and observe what will happen if Player 1 plays L1. In this subgame, Player 2 is constrained by rationality and will be forced to play R2. Consequently, Player 2's threat is not credible, and so (R1, L2) is ruled out as a subgame perfect equilibrium. In this example, only the equilibrium (L1, R2) is a subgame perfect equilibrium.

3. STRATEGY MAPPINGS AND RASMUSSEN'S OUTDATED COMPUTER DISK GAME

To understand the idea of a strategy mapping, consider Rasmussen's (1989) example of a rivalry between two software companies. This game revolves around two computer companies trying to decide whether to adopt a 3.5- or 5.0-inch format for recordable computer disks. Both companies are trying to coordinate so that if one produces a large disk, the other will also produce a large disk, and if one produces a small disk, the other will do the same. The normal form is depicted in Table 9.2. Notice that both players receive their highest payoff when they coordinate on the large disk, with both getting a payoff of 2. If they both select the small disk, then they both get a payoff of 1. In the case in which they do not coordinate, both players get a payoff of −1.

Now let us consider the extensive form of this game, displayed in Figure 9.3, where Player 1 moves first. Let us ignore the payoffs for now and just focus on the players' strategies. Player 2 has four strategy combinations that represent his choices (large, small) in conjunction with Player 1's choices (LARGE, SMALL). First, notice that if Player 1 picks LARGE, Player 2 can play either large or small, and if Player 1 picks SMALL, Player 2 can play either large or small.

TABLE 9.2 NORMAL FORM OF OUTDATED COMPUT-ER DISK GAME

		Player 2	
		Large	Small
Player 1	Large	2, 2*	−1, −1
	Small	−1, −1	1, 1*

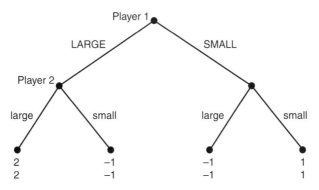

FIGURE 9.3 Extensive form of outdated computer disk game

One strategy that Player 2 can adopt is to play large if Player 1 picks either LARGE or SMALL. This is the (large, large) strategy, in which Player 2 responds to whatever Player 1 picks by playing large. Another strategy is the (small, small) strategy, in which Player 2 responds to whatever Player 1 picks by playing small. Then there is the (large, small) strategy, in which Player 2 plays large if Player 1 picks LARGE and plays small if Player 1 picks SMALL. Finally, there is the (small, large) strategy, in which Player 2 plays small if Player 1 picks LARGE and plays large if Player 1 selects SMALL. Now we have a complete mapping of the strategies that Player 2 can play, contingent on what Player 1 selects.

Using behavioral strategies, we can reconstruct the normal form to reflect all of the strategy combinations that are possible using this extensive form. The normal form game in Table 9.3 reflects the expanded strategy set, listing all of Player 2's strategies and their associated payoffs for each player.

Note that in Column 1, if Player 1 selects LARGE, Player 2 selects large (2, 2), and if Player 1 selects SMALL, Player 2 selects large (–l, –l). In Column 2, if Player 1 selects LARGE, Player 2 selects large (2, 2), and if Player 1 selects SMALL, Player 2 selects small (1, 1). In Column 3, if Player 1 selects LARGE, Player 2 selects small (−1,−1), and if Player 1 selects SMALL, Player 2 selects large (−1, −1). Finally, in Column 4, if Player 1 selects LARGE, Player 2 selects small (−1,−1), and if Player 1 selects SMALL, Player 2 selects small (1,1).

There are three Nash equilibria to this expanded game, as denoted by the asterisks in the table: (LARGE, large, large), (LARGE, large, small), and (SMALL, small, small). However, only (LARGE, large, small) is subgame perfect. Why is this the case? Let us first consider the (LARGE, large, large) equilibrium. The extensive form depicted in Figure 9.4 traces the equilibrium path along the dashed and dotted line. This equilibrium requires that when Player 1 picks LARGE, Player 2 also plays large, and when Player 1 picks SMALL, Player 2 still plays

TABLE 9.3   EXPANDED STRATEGIES FOR THE OUTDATED COMPUTER DISK GAME AND EQUILIBRIA

		Player 2			
		large, large (1)	large, small (2)	small, large (3)	small, small (4)
Player 1	LARGE	2, 2*	2, 2*	−1, −1	−1, −1
	SMALL	−1, −1	1, 1	−1,−1	1, 1*

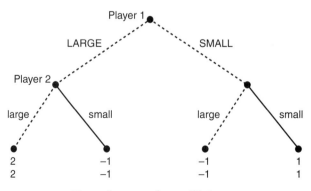

FIGURE 9.4   Not a subgame perfect equilibrium

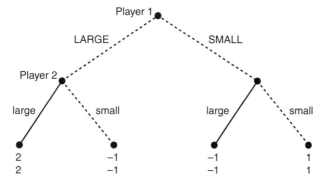

FIGURE 9.5 Not a subgame perfect equilibrium

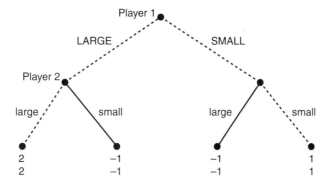

FIGURE 9.6 A subgame perfect equilibrium

large. However, if Player 1 picks SMALL, it would be irrational for Player 2 to play large, since his payoff would be −1 as opposed to 1.

Now consider Figure 9.5, which depicts the (SMALL, small, small) equilibrium that calls for Player 2 to respond to whatever strategy Player 1 selects by playing small. In this case, if Player 1 selects SMALL, Player 2 would respond with small. However, if Player 1 plays LARGE, it would not make sense for Player 2 to respond with small, since he could do better by playing large and getting a payoff of 2 instead of −1.

Hence, the only subgame perfect equilibrium for this game is the (LARGE, large, small) equilibrium depicted in Figure 9.6, in which Player 2 does best by responding to LARGE with large and to SMALL with small.

4. PLAYER 1 MOVES TWICE GAME

Consider another game, depicted in Figure 9.7. In this game, Player 1 moves first, Player 2 moves second, and Player 1 takes the final move.

This game has three subgames: the whole game, the game starting at Player 2's node, and the game starting at Player's 1 second move. Before we generate a mapping of strategies, notice that if Player 1 plays her R1 strategy on the first move, then Player 2 has no move. However, we still have to account for what Player 2 will do if Player 1 selects R1. In the outdated computer disks game, all players had a choice at their decision node, so all choices were on the equilibrium path. In this game, if Player 1 selects R1 the game terminates, but Player 2 must also consider this first move when considering all the different strategy combinations. Although Player 2 has no choice to make if R1 is selected, we still have to account for that choice.

Player 2 only has two strategies, A and B, but Player 1 has four strategies: (R1, R2), (R1, L2), (L1, R2), and (L1, L2). Let us begin with the cases in which Player 1 has the first move and selects R1, (R1, R2) and (R1, L2). In these cases, Player 2 does not get a move, so Player 1 will get a payoff of 0 and Player 2 will get a payoff of 3. Now, when Player 1 selects L1 and Player 2 selects A, the game ends, so regardless of Player 1's final move of R2 or L2, Player 1 will get 1 and Player 2 will get −1. When Player 1 selects L1 and Player 2 selects B, if Player 1 then plays R2, she will receive a payoff of 0 and Player 2 will receive a payoff of 2, but if she then selects L2, both players will get a payoff of −2. The strategy mapping and payoffs for this game are contained in Table 9.4.

There are four Nash equilibria in this game: (R1, R2, B), (R1, L2, B), (L1, R2, B), and (L1, L2, A). Notice in Figure 9.7 that if Player 1 gets a second move she will always select R2, so any subgame equilibria must have R2 as an element. Equilibria containing the strategy L2 can

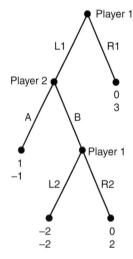

FIGURE 9.7 Player 1 moves twice game

TABLE 9.4 STRATEGY MAPPING AND PAYOFFS

| | | Player 2 | |
		A	B
Player 1	R1, R2	0, 3	0, 3*
	R1, L2	0, 3	0, 3*
	L1, R2	1, −1	0, 2*
	L1, L2	1, −1*	−2, −2

also be ruled out, since Player 1 will never select this strategy. Given these guidelines, we can rule out (R1, L2, B) and (L1, L2, A) as candidates for subgame perfect equilibria. We are thus left with (R1, R2, B) and (L1, R2, B), and if we examine the equilibrium path on the extensive form, we can see that neither can be ruled out and both are subgame perfect equilibria.

Both of the remaining equilibria are perhaps frustrating to Player 1, because if she gives a move to Player 2, then each equilibrium requires Player 2 to select B on his turn and not the A strategy that Player 1 desires. The only way that Player 1 can achieve her best payoff is to somehow convince Player 2 to select A when she plays R1. But according to our rationality assumption, Player 2 will never be convinced, because if he selects A he will receive a payoff of −1.

5. KREPS AND WILSON'S UP-DOWN GAME

Now let us examine another game called the up-down game, presented in Figure 9.8 (Kreps and Wilson 1982, 869). In this game, Player 1 has the option of playing Down, in which case the game ends, or playing Up, in which case Player 2 can move either Left or Right.

The strategy mapping and payoffs of this game are presented in Table 9.5. Again, when Player 1 selects Down, Player 2 has no choice and both players receive a payoff of (1,1). This game has two Nash equilibria, which are marked with asterisks in the table. Now we have to determine if both equilibria make sense rationally.

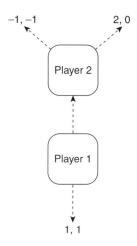

FIGURE 9.8 Up-down game

TABLE 9.5 UP-DOWN GAME STRATEGY MAPPING AND PAYOFFS

		Player 2	
		Left	Right
Player 1	Up	−1, −1	2, 0*
	Down	1, 1*	1, 1

Does the (Down, Left) equilibrium make sense? No, because given a move, Player 2 will always select Right, and, as in the previous game, Player 2 wants Player 1 to select Down. Player 2 can only threaten Player 1 that he will play Left, producing a (−1, −1) payoff, if Player 1 does not play Down, but this is a noncredible threat. Consequently, (Up, Right) is the subgame perfect equilibrium.

C. SUBGAME PERFECT EQUILIBRIUM AND THE NEED FOR REFINEMENT

Subgame perfect equilibrium is a refinement of Nash equilibrium because it requires players to map out a complete plan of action in the extensive form so that we can consider actions both on and off the equilibrium path. This method has allowed us to eliminate equilibria that are supported by irrational game play, because the games we have examined assume complete and perfect information (i.e., each player's decision node was a singleton). This assumption allowed us to work backward through the game tree, forcing players to consider the implications of carrying out a futile threat. Once we eliminated those empty threats that supported an equilibrium, the subgame perfect solution to the game remained.

When we examine games of imperfect and incomplete information in the next chapter, we will be concerned with games in which players face multiple nodes without knowing for sure the node (or information set) at which they are located when they make a decision. While subgame perfect equilibrium can be used in these sorts of games, it is often unable to evaluate irrational equilibria as a result of a game's information structure. Subgame perfect equilibrium does not capture the idea that behavior should be rational every time a player confronts multiple nodes. Its power lies in the fact that each player has a single decision node, allowing all subgames to be evaluated.

D. CENTIPEDE GAME

1. HOW THE CENTIPEDE GAME IS PLAYED

The centipede game offers a dilemma in that backwards induction makes a unique subgame perfect equilibrium prediction but experimental subjects often deviate from that solution (Rosenthal 1981). This game is often used to represent coalition-building behavior in legislative bodies, where individual members of a coalition interact over time and reciprocity behavior becomes important. In this game, each coalition member is better off if he can trust another coalition member, but each member has incentive to not trust other members. Figure 9.9 presents a four-move game in which two players have two options to either take a sum of money or pass the play to the other player. Players have an incentive to pass the play, since both players can receive higher payoffs as game play progresses. Once one player takes the money the game is over. Notice that the payoff structure of the game gives Player 1 the option of taking the money or passing the play to Player 2. If Player 1 passes and Player 2 takes the money, then Player 1 gets a lower payoff than if she took the money on the first move (0.20 as opposed to 0.40). If Player 2 decides to not take the money and passes the play to Player 1 and Player 1 takes the money, she gets 0.40 as opposed to 0.80. If Player 1 decides to not take the money and again passes the play to

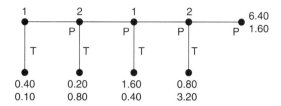

FIGURE 9.9 Centipede game

Player 2 and Player 2 takes the money, Player 1 gets 0.80 as opposed to 1.60. On the last round, Player 2 should take the money and get 3.20 as opposed to 1.60.

Now let us solve for the equilibrium using backward induction. Starting at the last node, Player 2 must choose between 1.60 and 3.20, so he selects 3.20; then Player 1 has a choice between 0.80 and 1.60 and so selects 1.60; then Player 2 has a choice between 0.40 and 0.80 and selects 0.80; and finally Player 1 has a choice between 0.20 and 0.40 and selects 0.40. Hence, the subgame perfect equilibrium is for Player 1 to take the money on the first move.

When McKelvey and Palfrey conducted experiments on this game, they discovered some odd outcomes:

> [W]e designed and conducted an experiment, not to test any particular theory (as both of us had been accustomed to doing), but simply to find out what would happen.
>
> However, after looking at the data, there was a problem. Everything happened! Some players passed all the time, some grabbed the big pile at their first opportunity, and others seemed to be unpredictable, almost random. But there were clear patterns in the average behavior, the main pattern being that the probability of taking increased as the piles grew. (Palfrey 2007, 426)

Only 37 of 662 games in this experiment ended on the first move, and players in 23 experiments passed at every move (McKelvey and Palfrey 1992). Considering the raw data, subjects did not even come close to the subgame perfect equilibrium prediction. McKelvey and Palfrey (1992, 805) comment:

> This is equivalent to asserting that subjects did not believe that the utility functions we attempted to induce are the same as the utility functions that all subjects really use for making their decisions. I.e., subjects have their own personal beliefs about parameters of the experimental design that are at odds with those of the experimental design.

2. CENTIPEDE, REPUTATIONS, AND QUANTAL RESPONSE EQUILIBRIUM

McKelvey and Palfrey (1992) conjectured that reputational effects might be an explanation for players' low level of compliance. If a player builds a reputation of being an altruist—a person who always passes—then experimental outcomes can be more easily explained in terms of the model. If the first player knows that she is playing against an altruistic person, then she should pass more readily to achieve a larger payoff. The assumption that a subject believes that there are a large number of other subjects who are altruistic explains a large amount of variation in the data. In effect, subjects had utility for money, but they also had other-regarding preferences and beliefs that affected their behavior.

McKelvey and Palfrey (1992) conducted a series of repeated versions of this game that allowed them to measure the number of altruistic players involved. In their analysis, they estimated a quantal response equilibrium, which we discussed in Chapter 7. This approach took into account the fact that players had beliefs about whether they were playing against an altruistic (nice) person or a selfish (mean) person. The researchers found that players playing against altruistic subjects were more likely to pass and subjects playing against selfish subjects were more likely to take the money early in the experiment. These results provided important evidence that emotions can impact behavior within this type of bargaining structure.

3. CENTIPEDE AND CHESS PLAYERS

Palacios-Huerta and Volij (2009) conducted experiments with a six-move centipede game and chess experts. As the last chapter noted, chess players are experts in backward induction, so their behavior should more closely mirror the subgame perfect equilibrium prediction. Subjects including Grandmasters, International Masters, Federation Masters, and lesser players were recruited from three international open chess tournaments. Students from undergraduate classes were also recruited. Students were paired against each other and chess experts were paired against each other. The paired subjects were placed in separate rooms at the tournament, and subjects text messaged their responses to each other. Subjects only played the game once.[1]

The responses of the student subjects were consistent with McKelvey and Palfrey's (1992) results, in that only a few subjects selected the subgame perfect equilibrium solution and most ended the game about halfway through the process (only 7.5 percent of student subjects ended the game at the initial node). However, analysis of chess players playing against chess players revealed a different story. The experts were 10 times more likely to end the game near the subgame perfect equilibrium prediction. The Grandmasters had a perfect record, with all ending the game at the initial node, whereas approximately 70 percent of the International Masters, Federation Masters, and other chess players ended the game at the initial node. The data clearly showed that chess players were able to discern the equilibrium and therefore ended the game significantly sooner than did student subjects. Consequently, by figuring out the game, the "experts" earned less money. What is important about these results is that they show that people's achievement of the equilibrium prediction depends on their cognitive ability.

In this section, we examined the centipede game using two different types of experiments. Both experiments were conducted in different environments, so the comparison of results offers an example of increasing the model's external validity. One experiment showed that altruistic behavior impacts equilibrium predictions, while the other showed that cognitive ability impacts equilibrium predictions. These experiments then provide different and important insights into the behavior of people who interact in this type of bargaining environment.

E. ULTIMATUM BARGAINING GAMES

1. ULTIMATUM BARGAINING AND PROBLEMS WITH SUBGAME PERFECT EQUILIBRIUM

Recall the ultimatum bargaining game discussed in Chapters 1 and 2 in which two subjects have to decide how to divide up a sum of money. The first player gets to decide on the allocation and the second player gets to decide whether to accept or reject the offered amount. If the second player rejects the offer, then neither player receives a payoff, but if the offer is accepted, then the players receive the amount in the offer. The game is depicted in Figure 9.10.

Say that the players are bargaining over six dollars. Player 1 can make an unfair offer of giving Player 2 one dollar and keeping five dollars, or Player 1 can make a fair offer of a 50/50 split of the money. The backward induction outcome is easy to observe, and if Player 1 strictly values expected utility, she should make the unfair offer and Player 2 should accept it.

However, as in the centipede game, the subgame perfect equilibrium prediction is often not realized, since subjects have inequality-averse and fairness preferences. The average outcomes in these experiments tend to be around 50 percent, or the first player plays the fair strategy more often (Roth 1995a; Camerer and Thaler 1995). This type of division also occurs in dictator games, in which the first player makes an offer and the responder is not given a choice. In these games, the dictator should just propose a 100/0 split and get all of the money, but other outcomes (such as fair division) also prevail.[2]

2. ULTIMATUM BARGAINING AND COMMUNICATION

Many factors can influence the outcome in ultimatum games, such as whether bargainers play anonymously or face-to-face, and if players are able to communicate with each other. Roth (1995a) reports on an ultimatum experiment in which he varied three treatments. In Treatment 1, subjects played the game anonymously without communication. In Treatment 2, subjects had two minutes to discuss the game before they played against each other. In Treatment 3, subjects had two minutes to discuss anything except the game. This treatment

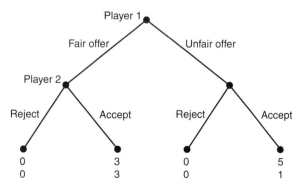

FIGURE 9.10 Ultimatum game

was used to test the hypothesis that if subjects were to get to know each other on a social level, then they would be more willing to play fair. Roth found that 33 percent of subjects in the anonymous Treatment 1 selected the unfair strategy, 4 percent of subjects in Treatment 2 selected the unfair strategy, and 6 percent of subjects in Treatment 3 selected the unfair strategy. What these experiments show is that players' ability to communicate or get to know their opponent makes a difference in this type of game. That is, players are able to display fair reciprocity behavior when they establish communication among themselves.

3. BARGAINING WITH SOCIAL PREFERENCES TURNED OFF

In another set of experiments conducted by Johnson et al. (2002), social preferences were effectively "turned off" by having subjects play against robots (or a computer program) that played the game with the goal of maximizing its own payoff each period. Johnson and colleagues reasoned that playing against a robot should eliminate human emotions such as a sense of fairness. Subjects played eight rounds. The payoff-maximizing equilibrium offer was $1.25 and subjects averaged $1.84. The researchers felt that subjects could do better with more experience, so they provided a brief tutorial on the concept of backward induction and then allowed the subjects to play the experiment again. In this new iteration, the experienced subjects came within pennies of the $1.25 equilibrium offer. What this result tells us is that, again, subjects need experience with backward induction reasoning and the typical utility function cannot accurately measure preferences in these types of experiments, since subjects appear to have social preferences.

4. ULTIMATUM BARGAINING AND CULTURAL EFFECTS

Another factor that might influence play in these types of bargaining experiments is cultural effects.[3] That is, considerations of fairness may vary from culture to culture. Henrich and colleagues (Henrich 2000; Henrich et al. 2001) conducted a collaborative project with anthropologists and economists in which they conducted ultimatum games in 13 countries from Africa, South America, and southeast Asia. The study found that the cultures that proposed the most unfair offers were the Machiguenga of Peru, the Hadza of Tanzania, the Tsimané of Bolivia, the Quichua of Ecuador, the Torguud of Mongolia, and the Mapuche of Chile. Camerer (2003, 70) comments:

> The Machiguenga are quite socially disconnected. The economic unit is the family; families hunt, gather, and practice swidden ("slash and burn") manioc farming. Anonymous transactions within a village are rare. Their society is the opposite of the bar on the television series *Cheers* ("where everybody knows your name")—they don't have proper names for other Machiguenga except for relatives. Perhaps the extreme social and economic isolation of the Machiguenga explains why they have no sharing norm.

The cultures that proposed the fairest offers were the Ache headhunters of Paraguay and the Indonesian Whalehunters, whose offers were more than half of a payoff on average. Camerer notes that cultures who offered a lot did so because if they did not, there would be turmoil in the village. He observes (2003, 71–72):

The anthropologists think these hyperfair offers represent either a norm of oversharing because game caught in a hunt cannot be consumed privately, or a potlatch or competitive gift-giving. Accepting an unusually generous gift (such as excess meat caught in a successful hunt) incurs an obligation to repay even more, and is considered something of an insult (since it implies that the giver is a better hunter than the receiver). Hyperfair offers are often rejected, consistent with the competitive gift-giving interpretation. These offers are a reminder not only that self-interest is typically violated in these games, but also that offers and rejections are a language with nuance and cultural variation.

These important experiments show that cultural differences do matter in these types of bargaining situations. Emotions still impact outcomes, but these emotions can be interpreted differently from culture to culture.

5. PHYSICAL ATTRACTION AND ULTIMATUM BARGAINING

Consider an ultimatum bargaining game conducted by Solnick and Schweitzer (1999) that aimed to examine how physical attractiveness and gender impact bargaining. To establish a measure of attractiveness, 70 student subjects were photographed. The researchers used photos instead of having subjects observe live faces to control for the measure of attractiveness that they established. This measure was produced by a panel of 20 judges (nine women, 11 men), who subjectively rated the attractiveness of the photographed students. Of the 70 photos, 24 were selected and placed in one of the following categories: six most attractive men, six most attractive women, six least attractive men, six least attractive women. The photos were then placed in a book in random order. During the experiment, the photos in the book became subjects' opponents as they bargained over 10 dollars. Proposer subjects made offers to each photo and responder subjects specified minimum acceptance levels for each photo. Each subject participated in 24 rounds of the experiment, corresponding to the 24 photos. The results of the study found that attractive people earned 8 percent to 12 percent more than unattractive people. The biggest winners were men, who earned 13 percent to 17 percent more than women. The authors conclude (1999, 210):

> These findings are consistent with the "beauty premium" and persistent gender gap in the labor market. Attractive people and men earn more, but prior work has been unable to disentangle whether these populations are offered more, demand more, or both. Results from this work suggest that attractive people and men receive more largely because they are offered more.

One clear objection to this experiment would be the subjective measure of attractiveness. To provide a more objective measure, Zaatari, Palestist and Trivers (2009) have defined attractiveness in terms of facial symmetry (FA). They note (628):

> Symmetrical individuals (especially males) are more attractive to the opposite sex in a wide range of species, including humans...
>
> FA is a measure of biological quality because it measures an important underlying variable, the degree of developmental stability....The more symmetrical an organism is (low FA), the better is the rest of the phenotype. Symmetry is shown to have strong positive associations with immune strength and resistance to parasites, ... ability to escape predators, speed, strength, mental acuity and ability to cope with a wide range of stressors.

Zaatari, Palestist and Trivers (2009) took photos of 116 young adults and used software to measure the degree of FA in each individual's face. They then selected 20 photos and divided them into four equal groups of five subjects each: symmetric men, symmetric women, asymmetric men, and asymmetric women. As in Solnick and Schweitzer's (1999) experiment, subjects played against the photo. The results found that when people played against people with symmetric faces, they tended to give more. Zaatari, Palestist, and Trivers (2009, 630) comment:

> When asked why they gave one photo more than the other, subjects' responses revealed a striking dichotomy: 35 gave more to a photo because they thought it was more attractive or cute and of these, 29 picked the symmetrical photo. Eleven gave more to the photo they said needed it more, and all chose the asymmetrical photo.

These experiments provide evidence that in a bargaining situation, the appearance of an individual may impact the outcome of a decision.

F. TRUST GAMES

The last class of bargaining games that I want to discuss in this chapter is the trust game (Berg, Dickhaut, and McCabe 1995). In this game, one player is given an endowment of money, such as 10 dollars, and granted the first move. That player can keep any portion of the endowment or give any portion to a second player. If she decides to keep the entire 10 dollars, then the game is over, and Player 1 will get 10 dollars and Player 2 will get nothing. But if she decides to give part or all of the endowment to Player 2, then the amount she gives will be multiplied by a factor of three, and Player 2 will have to decide whether to give any money back to Player 1. If Player 1 trusts Player 2, then she should give him the whole endowment of 10 dollars, which will be converted to 30 dollars, assuming that Player 2 will give her back at least 10 dollars and preferably 15 dollars. However, the subgame perfect equilibrium of this game is for Player 1 to not trust Player 2 and to simply keep the 10 dollars and terminate the game.

Berg, Dickhaut, and McCabe (1995) conducted a double-blind experiment in which subjects did not know or see the other subjects against whom they were playing. Subjects were placed into two rooms A and B. Subjects in Room A had to decide how much to invest, and once that decision was made, the subject related the amount to the experimenter, who tripled the amount. This amount was disclosed to subjects in Room B, who then had to decide how much to return to the subjects in room A. The subgame perfect equilibrium was rejected 30 of 32 times. On average, subjects in Room A sent an investment of five dollars to Room B. Five of the 32 subjects sent their entire 10 dollars, while only two of the 32 subjects behaved as predicted and sent nothing. Room B behavior investments of five dollars yielded a return of $7.17 and investments of 10 dollars yielded a return of $10.20. About two-thirds of Room B subjects did not reciprocate and return the money. Although five-dollar investors received a significant return on their investment, 10 dollar investors would have been better off not sending an investment. Again, recall that the subgame perfect equilibrium of this game is for the first player to not trust the second player and keep the 10 dollars. The experiment's results disconfirmed that prediction. As in the ultimatum games described previously, variations in the trust game may be based on cultural differences (see Willinger, Lohmann, and Usunier

2003). These games show that when using rationality as a standard, subjects trust when they should not. Again, social preferences may help us to understand why this occurs and the ways in which trust can bias decisions.

G. NEUROECONOMICS

In this section I will introduce an area of research that is attempting to produce more refined measures of emotions by examining the neural network of the brain (i.e., the wiring in our brains that is composed of neurons). Neuroeconomics is the joining of neuroscience (the study of brain processes) with economics to study how the human brain makes decisions (neural networks) when emotions are involved (see Camerer, Loewenstein, and Rabin 2003 for a review). This method is premised on the idea that different locations of the brain, or clusters of neurons at different locations, represent different emotions and these emotions can be identified by locating the area of the brain where the neurons are activated. Assuming that we have a precise map of the brain that indicates precisely which emotion is associated with the location of neurons, it should be possible to identify the neurons that are activated when an emotion is activated, and to identify the emotion that is sparked when a decision is made. In short, this area of study attempts to measure levels of brain activation within a region of the brain to determine what emotion may be involved when a decision is made. This method attempts to read (or map) a subject's brain as he or she plays these bargaining games. These types of experiments provide a good testing ground, since specific emotions can be induced in a relatively controlled environment where it is possible to isolate (although not precisely) the location of the resulting emotion or kinds of emotions.

DeQuervain et al. (2004) used a positron emission tomography (PET) imaging of a subject's brain to examine the emotional effect of punishment in a trust game. The game was modified so that the first mover was given an option to punish the second player, for a cost, when this player was not truthful. Punishment costs would reduce the payoff of both players. The purpose of the experiment was to discover what happens to the first mover's brain when the second player is not trustworthy and the first mover punishes him. The authors identify an area in the midbrain called the dorsal striatum where reward stimuli from decisions are processed. As a result, it should be possible to examine whether the first player derives satisfaction from the act of punishing the second player by observing increased activation of the dorsal striatum. The experiment showed that subjects did gain satisfaction from the act of punishing, which runs contrary to self-regarding preferences. Although costly to themselves, some subjects were willing to bear that cost to be able to punish the second player. This indicates that punishment (revenge) is valued in addition to monetary rewards.

Sanfey et al. (2003) used functional magnetic resonance imaging (fMRI) to take pictures of brain activity to study whether rates of unfair offers increased activation in the dorsolateral prefrontal cortex, an area of the brain associated with negative emotions such as pain, thirst, and hunger. Their results showed that unfair offers did increase activation in that portion of the brain, indicating a strong emotional response. However, it was unclear what emotion was responsible for the response.

Finally, in a different approach, some researchers have become interested in how hormones affect human emotions. Neuropeptide oxytocin (OT) is a hormone that supposedly induces pro-social behavior in animals. To test this hypothesis on humans, Kosfeld et al. (2005) conducted an experiment in which they gave subjects the hormone and let them play a trust game. This group was compared against a control group, which played a risk game. The experiment showed that the players who made the first offer and were given OT were significantly (20 percent) more trusting than players in the control group.

These simple games help us understand how hormones affect decision-making. The problem that this approach must overcome is our inability to precisely identify those regions of the brain where the neurons are sparked. At present, only broad estimates can be made. For instance, the locations of neurons that spark emotions are difficult to identify because we do not have an accurate map of the brain. Kaku (2009, 86) comments:

> Some scientists have advocated a "neuron-mapping project," similar to the Human Genome Project, which mapped out all the genes in the human genome. A neuron-mapping project would locate every single neuron in the human brain and create a 3-D map showing all their connections. It would be a truly monumental project, since there are over 100 billion neurons in the brain, and each neuron is connected to thousands of other neurons. Assuming that such a project is accomplished, one could conceivably map out how certain thoughts stimulate certain neural pathways. . . . Thus one would achieve a one-to-one correspondence between a specific thought, its MRI expression, and the specific neurons that fire to create that thought in the brain.

While neuroeconomics has potential for helping us understand how emotions are invoked, we are still a long way from achieving the goals described by Kaku.

H. WAIT A MINUTE, ARE THESE REALLY SOCIAL PREFERENCES?

1. MANUFACTURED SOCIAL PREFERENCES

Although the experiments I have presented in this chapter indicate that subjects have a preference for money, there is another component to the utility function that I have been calling social preferences or pro-social behavior, such as altruism, fairness, or trust. But some scholars argue that it is not natural pro-social behavior that we see in these experiments; rather, it is behavior that has been manufactured in the laboratory, since the laboratory environment forces people into unnatural positions (Levitt and List 2007). Consequently, subjects behave in ways that they would not if the stakes were higher or if they did not know they were being observed. Levitt and List present several experimental results that do not match similar findings from the field, demonstrating that cooperative behavior may be found in the lab, but people in similar real-world situations may not cooperate. Hence, they argue that although experiments find this additional component to a person's utility function, the type of behavior that composes it is unclear.

2. STRATEGIC IGNORANCE

An alternative theory argues that this "made-up" behavior can be explained by the idea of strategic ignorance, which says that given a choice of information, people prefer to avoid that information that could impact their beliefs and in turn their behavior (i.e., people prefer to ignore the consequences of their actions on others). According to the theory of strategic ignorance, subjects in experiments feel compelled to engage in pro-social behavior, although they would avoid this sort of situation given a "real" choice. Dana, Cain, and Dawes (2006) conducted an ultimatum game experiment in which a dictator was given 10 dollars to divide between herself and an opponent. The dictator was given the option of making an offer or leaving the room. If the subject decided to leave the room, she would get nine dollars and her opponent would receive nothing. About one-third of the subjects in this experiment chose to leave the room and "unfairly" leave the subject with nothing. The authors argue that if the subjects had altruistic intentions, they would at least propose a 9/1 allocation and not leave the room. In addition, if the subjects behaved rationally, then the dictator would have proposed a 10/0 allocation and maximized utility. Instead, the researchers observed that many subjects chose to remain ignorant of how their actions impacted others, diminishing altruistic behavior. They argue that this type of behavior is very different from the behavior interpreted as pro-social in other experiments.

I. SUMMARY

In this chapter, we examined the concept of subgame perfect equilibrium. This equilibrium concept allows us to refine the Nash equilibrium for extensive form games. The Nash concept was originally developed for normal form simultaneous games, but when extensive form games were introduced, the assumption of rationality was not always followed in certain situations. Subgame perfect equilibrium allows us to get rid of equilibria that are supported by noncredible threats.

The experiments in this chapter showed that we can identify subgame perfect equilibrium by using backward induction, but that equilibrium is not always evident in experiments. Bargaining experiments show that subjects have preferences for money but also exhibit social preferences, in that they care about the payoffs or welfare of other subjects. Many models have been created to explain behavior, some of which are based on emotions and others on nonemotions, such as the strategic ignorance theory. Studying emotions is difficult, but simple bargaining games can make this task manageable by observing the effects of emotions on decision-making. Future advances from neuroeconomics will surely provide more accurate estimates.

NOTES

1. The authors also conducted laboratory experiments in which students and chess players were paired against each other. In this treatment, subjects played the game 10 times.
2. This game was first proposed by Kahneman, Knetsch, and Thaler (1986).
3. The first paper that reported research on cross-cultural ultimatum games was Roth et al. (1991), who conducted these games in America, Israel, Japan, and Yugoslavia. See also Whitt and Wilson (2007) for an example of dictator experiments conducted in postwar Bosnia.

CHAPTER 10

IMPERFECT AND INCOMPLETE INFORMATION GAMES

A. THE STRUCTURE OF IMPERFECT AND INCOMPLETE INFORMATION GAMES

1. INFATUATION AND FICKLE GAMES

Consider a game in which two players, Igor and Olga, are friends and are deciding whether to go to a dance club or a coffee shop. Igor does not like talking very much and prefers to go to a dance club where the music is loud and he does not have to engage in conversation. Olga likes to talk, so she prefers a relatively quiet coffee shop to a loud dance club. The problem in this game is that Igor is crazy in love (or infatuated) with Olga. Igor wants to be with Olga regardless of whether or not she wants to be with him. Olga, on the other hand, is fickle: she likes Igor sometimes and wants to be with him, but at other times she does not like him and does not want to be with him, and these feelings are unpredictable. Let us imagine that there are two states of nature, A and B (or two different games that can be played). In State A Olga likes Igor, and in State B Olga does not like Igor. Tables 10.1 and 10.2 present two different normal form games that represent the two different states A and B.

Notice that State A is just like the battle of the sexes game discussed in Chapter 6. In this state, both players want to be together as opposed to apart, and Igor gets higher utility from

TABLE 10.1   STATE A: OLGA LIKES IGOR

		Olga	
		Dance	Coffee
Igor	Dance	2, 1	0, 0
	Coffee	0, 0	1, 2

TABLE 10.2   STATE B: OLGA DOES NOT LIKE IGOR

		Olga	
		Dance	Coffee
Igor	Dance	2, 0	0, 2
	Coffee	0, 1	1, 0

dancing than from talking and drinking coffee (with the opposite holding for Olga). In this game, Olga likes Igor and wants to connect with him. In State B, Igor also gets his highest utility from going dancing together and his second highest utility from drinking coffee together. But in this state, Olga does not like Igor and does not want to be with him, so she gets her highest utility by drinking coffee while he is dancing and her second highest utility by dancing while he is drinking coffee. In short, in this second game, Olga prefers to avoid Igor.

In State A, both players are trying to coordinate, but in State B, one player is trying to avoid another player. Now let us assume that Igor knows the state in which they are (i.e., he knows which game he is playing). Igor's behavior and strategy choices are constant across both games, so the state does not matter to him; he is infatuated with Olga in both states and wants to connect with her regardless of her feelings. However, Olga's behavior and strategy choices do vary by state (in State A she wants to connect with Igor, and in State B she wants to avoid him). If we assume that the probability of each state occurring is 50 percent, then we have a game of incomplete information. Olga does not know which state she is in when she makes a decision (i.e., she does not know if she likes Igor or not, so she does not know if she should avoid him or not). Without any more structure to this game, it is unclear how Olga will choose strategies other than randomly (which she does not want to do, since she has clear but divergent preferences in both states).

Games like this are referred to as games of incomplete information, since one player is privy to some information in the game structure (knowledge of the state of nature) and another player is not. To model this knowledge in the game structure, we must model a player's beliefs about the structure of the game. In this example, we have to assume that Olga has some beliefs about the state of nature she is in. That is, in order to make a decision, Olga needs to believe that she is in either State A or State B, since the game she plays has different payoffs in both states. As we have already learned, strategy choice is aligned with payoffs. If Olga does not know the state, she does not know how her strategy choice relates to payoffs and thus cannot make a rational choice. By modeling Olga's beliefs about the state of nature in which she believes herself to be, we can specify her choice of strategies given those beliefs. I would like to emphasize that these beliefs could be wrong, but both players believe their beliefs at the time a decision is made and act accordingly.

2. DISNEY MOVIES AND INCOMPLETE AND IMPERFECT INFORMATION

This chapter discusses imperfect and incomplete information in extensive form games. An imperfect information game occurs when the move of one player in a two-player game is not observed by a second player (i.e., a player does not perfectly observe the actions of another player). An incomplete information game occurs when one player does not have information that another player has, and this uninformed player has to move before or after the informed player (i.e., one player must make a decision without knowing the information that another player possesses). Incomplete information games were first developed by Harsanyi (1967–68). Myerson (2004, 1823) states:

> In Harsanyi's advanced sense, a game has *incomplete information* if, at the beginning of
> the game, some players have incomplete information about what other players know or

believe....In Harsanyi's...games, the players' different information is described by a collection of random variables, called the players' types, each of which is the private information of one player.

Consequently, private information about a player's type is essential to modeling incomplete information games. This information structure is also referred to as asymmetric information, or a situation in which one player has private information to which other players are not privy, such as a case in which one player knows her type but an opponent is unaware of the type of player against whom he or she is playing.

Incomplete information extensive form games are dynamic games that allow us to model instances in which a player has private information that is not observed by other players and this private information dictates game play. Consider the good or evil distinction. A player may know that she is evil, but her opponent may not. This information is important, because it could alter her opponent's strategy choice when the time comes to make a decision.

Disney movie studios have capitalized on a storyline in which one character has private information, usually involving evil intentions, that another character does not. For example, consider *Snow White*, in which the witch tricks Snow White into eating a poisoned apple. In *Aladdin* the evil Jafar disguises himself as an elderly man to trick Aladdin into retrieving a magic lamp that contains a genie. In *Cinderella*, the prince does not know that the evil stepmother is trying to hide Cinderella's identity. In *Beauty and the Beast*, Belle does not know if the Beast is good or bad. In *The Little Mermaid*, Ariel does not know that the sea witch Ursula wants to take her voice because she is evil and plans to use it to get the crown and trident from Ariel's father. In *101 Dalmatians*, the owners of the puppies do not realize that Cruella is evil and wants to make a coat out of the puppies. In *Toy Story*, the other toys think that Woody is evil and wrongly believe that he has gotten rid of Buzz Lightyear out of jealously. The list of movies that use this information structure goes on and on. Why is incomplete information incorporated into movie scripts? The answer is simple: it makes for great drama. The character who is not privy to the private information of another player has to make a decision based on beliefs that this player has at the moment of the decision. So if Snow White believed that the apple was from an evil witch, she would not eat it, but because she does not know this information, she does eat it. This assumption of incomplete information also allows game theorists to create drama within their own games.

In the last chapter, we examined extensive form games in which players knew where they were in the tree before they made their decisions. This complete and perfect information structure allowed us to use backward induction to solve for the subgame perfect equilibrium. However, when we examine extensive form games with imperfect and incomplete information, then the solution becomes a little more complicated.

In the next section, I will illustrate how imperfect information is incorporated into extensive form games. In the following section, I will introduce incomplete information and the notion of types, or chance moves. In the fourth section, I will introduce the concept of sequential rationality, which is a way to solve imperfect and incomplete information games and is another refinement of both the Nash equilibrium and subgame perfect equilibrium (Kreps and Wilson 1982). In the fifth section, I will examine signaling games and then outline a sender-receiver experiment on lying aversion. Finally, I will look at a persuasion experiment in the last section.

B. THE STRUCTURE OF INCOMPLETE INFORMATION IN GAME TREES

1. MATCHING PENNIES AND INFORMATION SETS

Incomplete information is modeled in a game tree by specifying information sets. To illustrate the notion of information sets, consider the matching pennies game. In this game, you and a friend match or mismatch pennies. If the coins match (either both tails or both heads), you pay your friend one dollar, and if they do not match, your friend pays you one dollar. Table 10.3 displays the normal form representation, and Figure 10.1 displays the extensive form.

The dashed line that connects the two decision nodes is referred to as an information set and indicates that Player 2 does not know which of the two strategies Player 1 has selected and therefore does not know the node at which she is located. Consequently, this extensive form is equivalent to the normal form, in which players must select strategies simultaneously.

Briefly, we can define an information set as a dashed line that connects two or more nodes. A player with incomplete and imperfect information only knows that game play has reached one of the nodes but is uncertain of the exact node. In this case, players may not know the node where they are located, but they do know that they are at either one node or another. I will discuss this idea in detail in section D of this chapter.

2. VARIED INFORMATION SETS IN A GUESSING GAME

Reconsider our follow-the-leader game described in Chapter 8. Players in that game had complete information, so there was no ambiguity about choices. Introducing various information sets connecting various nodes allows us to examine different types of information structures. Incomplete information converts the follow-the-leader game into a matching pennies–type game in which

TABLE 10.3 MATCHING PENNIES

		Friend	
		Heads	Tails
You	Heads	−1, 1	1, −1
	Tails	1, −1	−1, 1

Source: Gneezy 2005.

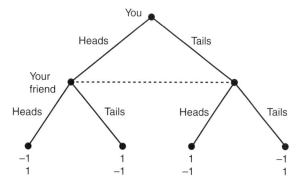

FIGURE 10.1 Matching pennies in extensive form

the specified information structure dictates which player has to guess. Recall that if all three players in the follow-the-leader game select L or R, then all three win a prize; otherwise, if there is a mismatch, they all lose (e.g., one player selects R and the others select L). To focus on the information structure of the examples in this section, I have not listed players' payoffs under the terminal nodes. Remember, however, that if they win they all get a payoff of 1, and if they lose they receive nothing. We will solve these types of games with payoffs later in the chapter. A game in which all three players have incomplete information is illustrated in Figure 10.2.

In this game, the information set connects both of Player 2's decision nodes, so she does not know if Player 1 has selected L or R. The information set also connects all of Player 3's decision nodes, so he does not know whether Player 1 or Player 2 has selected L or R. This game is just a simultaneous move game in which all players are uninformed about the choices that other players make. In this case, all three players must guess the other players' intentions.

Now consider the game in Figure 10.3 and notice that for Player 3 the information set does not connect all decision nodes, which indicates that the game is a sequential move

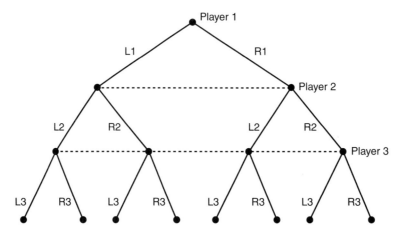

FIGURE 10.2 Three-player simultaneous game

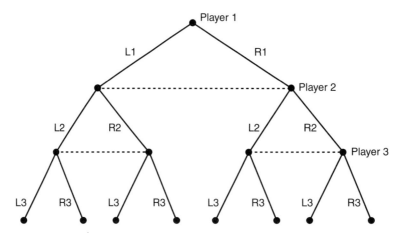

FIGURE 10.3 Player 3 does not know how Player 2 has moved

game (in which Player 1 moves first, Player 2 moves second, and Player 3 moves third). Since Player 1 moves first, she selects the initial strategy of L1 or R1. However, she must figure out (if possible) what Players 2 and 3 plan to do given her selection of L1 or R1.

This game represents the case in which Player 2 is uninformed about both whether Player 1 has selected L1 or R1 and, because he moves before Player 3, whether Player 3 is selecting L3 or R3. Player 3 knows if Player 1 has selected L1 or R1, but she does not know if Player 2 has selected L2 or R2. Player 3 only has to correctly guess Player 2's move, since she knows Player 1's selection.

The game in Figure 10.4 shows a case in which Player 2 knows which strategy Player 1 has selected but Player 3 is uninformed about the moves of Player 1 and Player 2. In this game, there is no information set for Player 2 and the information set for Player 3 extends across all of her decision nodes. Given this information structure, Player 2 knows Player 1's selection, so we can assume that he will follow suit and select the same strategy as Player 1 (either L or R). Now Player 3 faces a total guessing game. She knows that Player 2 will select the same strategy as Player 1, but this information is worthless since she still has to make a coin flip between L3 and R3.

Now consider Figure 10.5. In this game, Player 2 is uninformed about the strategy selected by Player 1, but if Player 1 selects L1, then Player 3 will be fully informed about all the choices before she has to make a decision. However, if Player 1 selects R1, Player 3 will know Player 1's move but will remain uninformed about Player 2's move. Player 2 is still left guessing, because he does not know what Player 1 has selected, although he does know what Player 3 will select. When Player 1 selects L1, Player 3 knows both this selection and Player 2's selection. If Player 3 sees that Player 2 has selected R2, then she knows they have already lost, but if she sees him select L2, she knows that she can seal the victory by selecting L3. When Player 1 selects R1, Player 3 knows this choice but does not know what Player 2 has played. In this case, Player 3 is left guessing only when Player 1 selects R1.

In Figure 10.6 the information set is curved so that we can model a different information structure. In this game, Player 2 is fully informed about Player 1's move. Player 3 does not know what Player 1 has selected, but she does know whether Player 2 has selected L2 (the

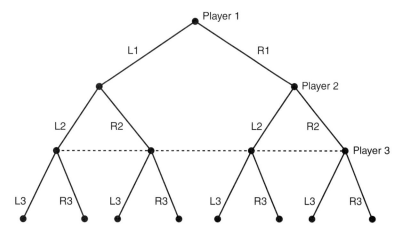

FIGURE 10.4 Only Player 3 has incomplete information

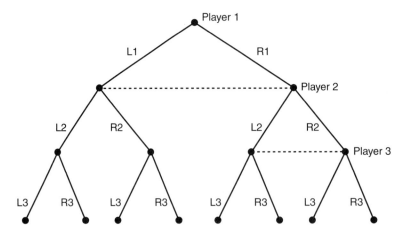

FIGURE 10.5   Player 3 has perfect information when Player 1 chooses L1 only

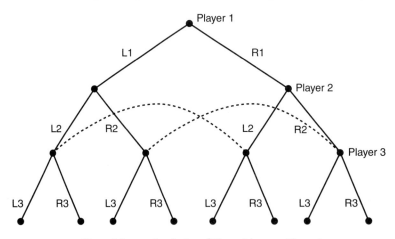

FIGURE 10.6   Player 3 knows the choice of Player 2 but not Player 1

information set extends to both of Player 3's L3 nodes) or R2 (the information set extends to both of Player 3's R3 nodes). Player 2 knows how Player 1 has selected, so we can assume that Player 2 will select the same strategy as Player 1. Hence, Player 3 only has to figure out what Player 2 selected, and because she knows whether Player 2 has played L2 or R2, she has perfect information and can win the game.

Finally, let us consider a four-player extensive form game with a more complicated information structure, depicted in Figure 10.7. At their choice nodes, what do Players 2, 3, and 4 know? Player 1 has two choices, L1 and R1. Player 2 does not know how Player 1 has selected, since both of his nodes are connected by an information set. Player 3 has four nodes, of which her R3 nodes are connected by an information set and her L3 nodes are not connected. This means that if Player 2 selects L2, then Player 3 knows the node at which she is located and therefore knows what Player 1 and Player 2 have selected. But if Player 2 selects R2, then Player 3 does not know what Player 1 has selected and thus does not know if

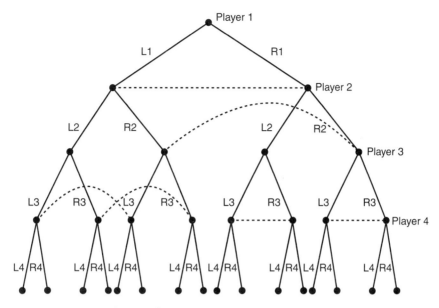

FIGURE 10.7 Complicated information sets

she is in the right or left side of the game tree. Finally, Player 4 has a curved information set that connects his R4 nodes and his L4 nodes on the left side of the tree and an information set that connects his R4 and his L4 nodes on the right side of the tree. Given this structure, if Player 1 selects R1, then Player 4 knows Player 1, Player 2, and Player 3's moves, but if L1 is selected, then Player 4 knows Player 3's move and not Player 2's move. This example illustrates that by using information sets within the extensive form framework, very complex information structures can be modeled.

3. RESTRICTIONS PLACED ON INFORMATION SETS

As Chapter 8 noted, perfect recall requires that players in these types of games never forget how they have moved. Let us look at a one-person game in which Player 1 has two consecutive moves but forgets how she moved on her first turn. A representation of this game is presented in Figure 10.8, in which the information set is connected to both of Player 1's choice nodes so that she does not know what move she made.

Consider also the two-person game depicted in Figure 10.9, in which Player 1 moves first, Player 2 moves second, and Player 1 has the final move. In this game, Player 1 selects L1 or R1, and Player 2 is aware of this choice. However, on her final move, Player 1 forgets her initial move. Most games assume perfect recall and do not allow this information structure.[1] The typical permitted game structure is depicted in Figure 10.10. In this game, Player 1 makes a move and Player 2 knows what move Player 1 has selected. However, on Player 1's final move, she knows what move she has made but does not know what move Player 2 has made.

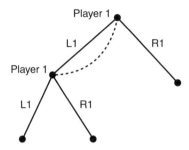

FIGURE 10.8 A one-player game in which the player forgets her move

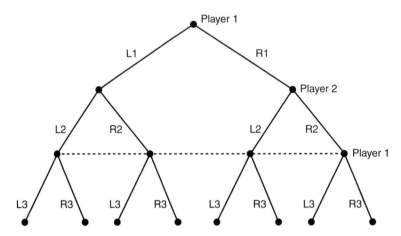

FIGURE 10.9 Player 1 forgets how she moved

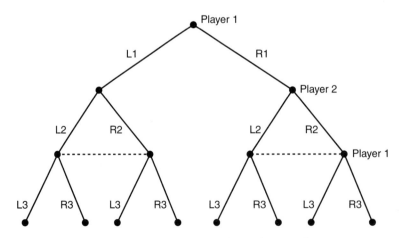

FIGURE 10.10 Player 1 does not know how Player 2 has moved

There are many more variations of how to model information sets in game trees. I will present another type of informational structure later in this chapter.

C. INCOMPLETE INFORMATION OVER PLAYER TYPES

John Harsanyi's (1967–68) seminal innovation was the introduction of explicit randomization in games by the incorporation of a random selection of a state of nature (or a chance move). This approach allows game theorists to assume a situation in which one player has observed a move "by nature," and although all players know the available moves, only certain players know the nature move (this is their private information). Private information models a situation in which some players know the state of nature and other players do not. The state of nature is a move determined by a probability distribution or a random device that game players cannot control.

Often the random state of nature is used as a proxy to represent different types. Instead of specifying State A or State B, we specify Type A or Type B, where a type can represent some quality such as performance level: for example, high or low. Consider football, which is played indoors and outdoors. Some players perform better or worse depending on where the game is played. We can specify that Type A players perform better outdoors and Type B players perform better indoors. After the type has been assigned, the player's payoff is impacted such that if he is a Type A player and the game is played indoors he will perform at a higher level (possibly achieving a higher payoff), and if he is a Type B player and plays outdoors he will perform at a lower level. When types are randomly assigned, a player knows her own type, but the player against whom she is playing does not have this information. A So in this example, a player knows that he is a Type A player who performs poorly indoors, but his opponent does not know this.

The random state of nature is usually the initial move and is represented as a hollow point. Some probability of occurrence is also associated with this node. If no probability is specified, it is assumed that the probability of occurrence is 50 percent (assuming there are two states of nature). Figure 10.11 presents a two-player game with two states of nature: good and bad. In this game, Player 1 knows if she is good or bad (i.e., she knows her type), but Player 2 does not know Player's 1 type.

In contrast, Figure 10.12 represents a situation in which Player 1 does not know her type (whether she is in a good or bad state of nature) but Player 2 does, although he does not know how Player 1 has moved.

Knowledge of the state of nature is relevant to a game, since a player might think that she is in a good environment when the opposite is true, leading her behavior to result in actions that are inconsistent with the assumptions of rationality. For example, two people might be playing a prisoner's dilemma game with communication in which the opponent sends a message that she is a cooperative player. In a good environment this would a truthful statement, but in a bad environment this would be a lie. A player makes an optimal move by relying on his or her beliefs, where beliefs are the probability distribution of being in a good or bad state of nature defined over the state of nature.

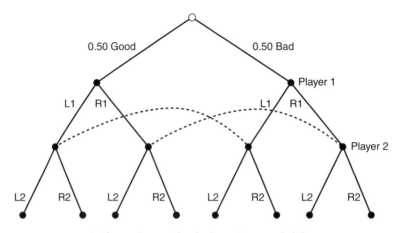

FIGURE 10.11 Is Player 1 in a good or bad state? 50% probability

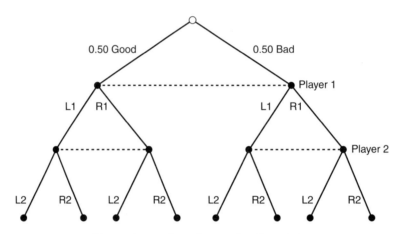

FIGURE 10.12 Player 1 does not know her type but Player 2 does

D. SEQUENTIAL RATIONALITY

1. ESTABLISHMENT OF BELIEFS AND RESTRICTIONS PLACED ON BELIEFS

A subgame perfect equilibrium allows us to determine if a player's move is rational for every subgame. This is possible in complete and perfect information games in which a player's information set is a singleton (i.e, players know the node at which they are located in every instance). That is, players confront a single node at every move of the game, so that when it is their turn to move, they know exactly where they are moving. With incomplete information in the extensive form, a player may not know the node at which she is located and may even confront multiple nodes, making it impossible for her to compute a subgame perfect equilibrium. The goal of this section is to solve for a sequential equilibrium (SE), which provides a solution to these types of games (Kreps and Wilson 1982). Kreps and Wilson's innovation is the notion that a player has sensible beliefs concerning the probability of being at a particular node. Hence, players make decisions about which strategy to select based on beliefs about their location in the game tree.

To solve games in which players confront multiple nodes (or do not know which move has been selected), we specify a player's "beliefs" as a probability distribution p (where $0 \geq p \geq 1$) over nodes in an information set, or as the conditional probability that a player is at the node where she thinks herself to be. For instance, a player may think that there is a 40 percent chance that she is on the right side of the game tree and a 60 percent chance that she is on the left side of the game tree. It is assumed that all players know the beliefs they have been assigned and are able to calculate expected utility based on those beliefs.

There are two requirements for an SE.[2] The first requirement says that at every information set, each player with a move has some belief about the node at which he or she is located that is conditional on having reached that information set—that is, players can calculate the relative likelihood of being at each information set. Put another way, this requirement simply means that at each information set, the player who has a move must have a belief system, or there is a probability distribution over nodes that players know. Players are required to have beliefs, but there is no stipulation about the reasonability of beliefs. The second requirement says that given beliefs on each information set, all strategies must be sequentially rational. This requirement simply means that in a choice between two alternatives where one choice yields a utility of 2 and the other yields a utility of 3, players must behave rationally and select the choice that leads to 3 rather than the choice that leads to 2. When put together, these requirements simply mean that players have beliefs and act rationally given these beliefs.

2. DERIVING A SEQUENTIAL EQUILIBRIUM

Consider the example in Figure 10.13 in which Player 2 does not know what move Player 1 has made. For illustrative purposes, let us assume that Player 2 believes that there is an 80 percent chance that he is on the right side of the game tree and a 20 percent chance that he is on the left side of the game tree.

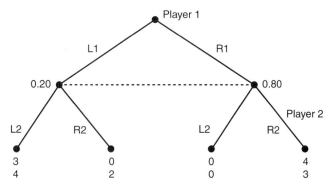

FIGURE 10.13 Sequential rationality

Sequential rationality means that a player should play optimally and maximize her expected utility at every information set. For instance, if Player 1 selects R1, Player 2 should select R2 as opposed to L2, since R2 yields 3 and L2 yields 0. Similarly, if Player 1 selects L1, Player 2 should select L2 and get 4, as opposed to selecting R2 and getting 2. These strategies have Player 2 selecting the largest payoff at each of his information sets. Now Player 1 must decide what strategy she should select, so she must calculate Player 2's expected utility to determine what he will select. Player 1 thus calculates Player 2's expected utility for selections of L2 and R2 as follows:

$$EU_2(L2) = 0.2\,(4) + 0.8\,(0) = 0.8$$
$$EU_2(R2) = 0.2\,(2) + 0.8\,(3) = 2.8$$

Now Player 1 knows that if she plays L1, Player 2 will play R2 and she will get 0. She also knows that if she plays R1, Player 2 will play R2 and she will get 4. Sequential rationality specifies that Player 1 will play R1 and Player 2 will play R2, resulting in their largest expected utility given their beliefs. The sequential rational equilibrium for this game is $(R1, R2, p = 0.8)$, or Player 1 selects R1 and Player 2 selects R2 when beliefs are 0.8.

Let us consider another example called the little horsey game[3] in which beliefs are *not* prespecified. In this case, Player 1 must calculate not only Player 2's expected utility, but also his beliefs. This game is depicted in Figure 10.14. Player 1 has three actions (L1, M1, R1), and, as before, Player 2 has a choice of L2 or R2. Player 1 can select R1 and end the game, in which case she gets a payoff of 1 and Player 2 gets a payoff of 3, or she can select L1 or M1 and give Player 2 a move. Whether she gives Player 2 a move depends on what she thinks

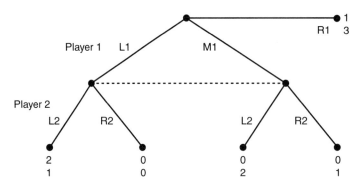

FIGURE 10.14 Little horsey game

TABLE 10.4 NORMAL FORM OF LITTLE HORSEY GAME

		Player 2	
		L2	R2
	L1	2, 1*	0, 0
Player 1	M1	0, 2	0, 1
	R1	0, 3	1, 3*

Player 2 will do. Player 1 will select R1 if she thinks that Player 2 will select R2 and give her a payoff of 0 instead of 1, but if she thinks that Player 2 will select L2, then she will select L1 and get a payoff of 2.

Before we calculate the SE, let us look at the normal form version of this game detailed in Table 10.4. Notice that this game has two pure strategy Nash equilibria, denoted by asterisks. In this game, Player 1 prefers the equilibrium (L1, L2) and Player 2 prefers the equilibrium (R1, R2).

To calculate the SE, we must first calculate Player 2's expected utility when Player 1 selects L1 and M1. Note that we do not need to calculate Player 2's expected utility for R1, because if Player 1 selects R1, Player 2 does not get a move.

When $0 \geq p \geq 1$:

$$EU_2(L2) = 1(p) + 2(1 - p) = 2 - p$$
$$EU_2(R2) = 0(p) + 1(1 - p) = 1 - p$$

Player 1 can observe that $(2 - p) > (1 - p)$, or $EU_2(L2) > EU_2(R2)$, so she knows that if she plays L1, Player 2 will play L2 and she will get a payoff of 2. She also knows that if she selects M1, Player 2 will select L2 and she will get a payoff of 0. Finally, she knows that if she selects R1, she will get a payoff of 1. Player 1 can deduce that selecting L1 yields her the highest payoff. The SE for this game is thus (L1, L2, $p \geq 0$).

E. SIGNALING GAMES

1. TRUTH-LYING GAME

In this section I will introduce a class of games referred to as signaling games. These games were devised by Michael Spence (1974), who won the Nobel Prize partly as a result of his introduction of this class of games.[4] I will discuss this type of game informally here and explain a reasonable way to solve it using very little math. We have already learned the math needed to solve signaling games formally using SE, and on the website I have included two examples of how to solve these types of games. I will first focus on the logic of the game. While we address this topic, also keep in mind the ways that signaling games underscore the difficulties of trying to model emotions in a game theory model.

To understand a simple signaling game, I would like to introduce a truth-lying game. This game has two players, a sender and a receiver. The sender delivers a message (in the form of a strategy choice) to a receiver, who takes an action based on the content of the message. The message can be delivered in many forms, but assume for now that it is a simple binary message that is dependent on the state of nature, so that it can be one of two types (e.g., a good or bad message). Only the sender knows the true state (or the true message), whereas the receiver can only observe the sender's message. As a result, the receiver has to make inferences about the true state from the sender's message. The sender (the informed player) has private information (her type) and makes the first move.[5] One problem inherent in these games is the potential for a conflict of interest between players, so the sender might want to deceive the receiver and send a dishonest signal. Another problem is that the sender and receiver may have a vested interest to cooperate, making the sender want to send an honest signal so that actions can be coordinated.

To clarify this situation, let us revisit our old friends Olga and Igor, and imagine that Igor has demanded to know Olga's true feelings toward him. But as we already know, Olga has a split personality in terms of her attraction to Igor; sometimes she likes him and sometimes she does not. Even Olga cannot understand why she feels this polarized attitude toward Igor, since she knows he is a nice but quirky guy. Because of her split personality, Olga is dreading her decision, since she does not know how she will feel about him when she gives him her answer. Olga is also smart and realizes that she does not have to be truthful and can either lie or tell the truth to Igor. But she also knows that whether or not she is truthful and what it means to be truthful depends on her state of mind (the state of nature) at the time that she blurts out her response. Let us call Olga's state of nature when she is altruistic and likes everyone including Igor S1, and her state of nature when she is an egoist and hates everyone including Igor S2. Consequently, her response to Igor will be determined partly by her state at the time she responds. As we know, Igor is infatuated with Olga, so he is going to have a hard time believing that she does not like him regardless of her attitude (or her state at the time of response).

Figure 10.15 presents a representation of this signaling game and is slightly different from the extensive form games we have seen thus far in this book. First, notice the two extensive form games, one at the top and another at the bottom, which are connected by three lines. In the top extensive form game, Olga (the sender) is an S1 type and must decide between telling the truth (left, top strategy) and lying (right, top strategy), while Igor (the receiver) must decide whether to believe Olga (B) or to not believe Olga (DB).[6] In the bottom game Olga is an S2 type and the game involves the same calculations as the top game. The solid line in the middle of the game represents a move by nature that determines Olga's (the sender's) type (S1 or S2) with some probability (in this example, 50%). The two dashed lines connecting the two extensive forms on each side mean that Igor (the receiver) does not know if he is in the top or bottom game when he makes his decision. Olga's payoff is listed first and Igor's second.

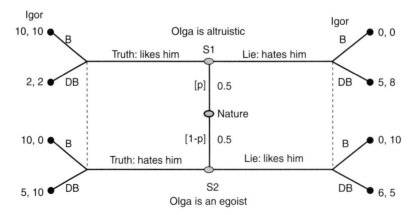

FIGURE 10.15 Lying-truth game

2. TRUTH-LYING AND GAMES OF CONFLICT AND COMMON INTEREST

The two games depicted in Figure 10.15 represent the difference between common and conflicting interest games. The top game is a common interest game, because both players have an incentive to take actions that will make them better off. In this state, Olga likes Igor and the best outcome is for her to tell the truth and confess that she likes him, and for Igor to believe that this is true, yielding both their best payoff of 10. Because this is the highest possible payoff for both players, it is in their common interest to take actions to achieve this outcome. The second best common outcome is for Olga to lie and tell Igor that she hates him, but for Igor to not believe her. In this case, Olga will feel guilty for lying, but at least Igor will not have believed her, yielding her a payoff of 5. For Igor, this situation is good, because he knows that Olga is lying about hating him, meaning that she actually likes him and yielding him a payoff of 8 (because this situation is not as good as her telling the truth and saying that she likes him). The third best common outcome is when Olga admits the truth about liking Igor and he does not believe her. This case leaves Olga disappointed, since she has finally revealed her true feelings, but they have been crushed by Igor's lack of belief. Igor is equally frustrated in this situation, because he really likes Olga and wants to hear her say those words he has been yearning to hear, but when he does hear them he cannot believe her. So both are disappointed and frustrated, but they can probably write this off as a big misunderstanding and still remain on the verge of a relationship, resulting in a payoff of 2 for both. The worst common outcome for both players is for Olga to lie and tell Igor that she hates him, and for him to believe it. Olga gets her lowest payoff of 0 from this outcome, since she feels guilty about lying and the fact that Igor now thinks she hates him when she really likes him. Igor feels equally horrible, since he believes that Olga hates him, so he also gets a payoff of 0. According to the payoff configuration, this is a game of common interest, since both players can avoid low payoffs by coordinating actions.

The bottom game represents a game of conflicting interests in which the state dictates that Olga hates Igor since she is an egoist. In this state, Olga and Igor's preferences do not align. Olga gets her highest payoff when she tells Igor the truth that she hates him and Igor believes her. In this case, Olga gains maximum satisfaction from hurting Igor's feelings and Igor is hurt in the worst possible way, so Olga gets a payoff of 10 and Igor gets a payoff of 0. Igor's best payoff in this game is when Olga lies and tells Igor that she likes him and he believes her. In this case, Igor gets a payoff of 10 and Olga gets a payoff of 0. When Olga lies and tells Igor that she likes him and he does not believe her, then Igor believes that Olga does in fact hate him. So although she had to lie, Olga got some of the results she wanted (since Igor knows she really hates him), so she gets a payoff of 6. Igor, on the other hand, is still in the picture, although he does not understand why Olga lied to him, so he gets a payoff of 5. Finally, when Olga tells the truth and informs Igor that she hates him and Igor does not believe her, then she gets some satisfaction from telling Igor her true feelings and thus gets a payoff of 5. Igor does not believe that she hates him but thinks that she really likes him, which nets his best payoff of 10. Therefore, according to this payoff configuration, both Olga and Igor must give the other player his or her

TABLE 10.5 GAME OF COMMON INTERESTS

		Igor	
		B	DB
Olga	Truth	10, 10	2, 2
	Lie	0, 0	5, 8

TABLE 10.6 GAME OF CONFLICTING INTERESTS

		Igor	
		B	DB
Olga	Truth	10, 0	5, 10
	Lie	0, 10	6, 5

worst outcome in order to achieve their best outcome, and so the players have conflicting interests.

Let us examine the normal form games of this extensive form, a practice that is useful (when possible) in fully understanding how a game will be solved. First, Table 10.5 presents the normal form of the top game, or S1.

There are two Nash equilibria in this game, (Truth, B) and (Lie, DB), with (Truth, B) being the Pareto optimal outcome. Table 10.6 presents the normal form of the bottom game of conflicting interest, S2.

There is no pure strategy Nash equilibrium in this game. However, notice that Igor has two best outcomes in response to Olga's truth-telling and lying strategies: (Truth, DB) and (Lie, B). This game also has a compromise position in which Olga lies and Igor does not believe her (6, 5). However, if both players are in this cell, then Igor has an incentive to defect, so there is no outcome in pure strategies.

3. CALCULATING A SEQUENTIAL EQUILIBRIUM FOR THE TRUTH-LYING GAME

Now that we have laid out the two games in their normal form, let us think about how to solve the game depicted in Figure 10.15. First, let us assume complete and perfect information (i.e., let us ignore Igor's information sets). In game S1, when Olga lies Igor does not believe her and they get a payoff of (5, 8); when she tells the truth Igor believes her and they get the highest payoff of (10, 10). Although the normal form game has two Nash equilibria, the sequential structure of this game eliminates the (5, 8) equilibrium, since if Olga knows she is in State 1 she will always tell the truth and get 10, as opposed to lying and only getting 5.

The S2 game is different, because when Olga tells the truth in this state, Igor will select DB and they will get a payoff of (5, 10), and when she lies in this state, he will select B and they will get a payoff of (0, 10). Notice that regardless of the strategies that Olga uses, Igor will get his best payoff of 10, so he will be indifferent to Olga's actions if he knows he is in S2. It is Olga who prefers the truth in this state, since telling the truth will get her a payoff of 5 as opposed to 0. The sequential nature of this decision creates an equilibrium in which Olga tells the truth and Igor does not believe her, yielding the (5, 10) payoff.

Although S1 has multiple equilibria and S2 has no pure strategy equilibrium, the sequential nature of choice provides a prediction in both cases. Now, however, let us reconsider the information set that prevents Igor from knowing the state. We know that in S1 Igor prefers to believe, and in S2 Igor prefers to not believe. But Igor also knows that in S1 Olga prefers to tell the truth that she likes him, and in S2 she prefers to tell him the truth that she hates him. Any equilibrium would have to support these beliefs. In both states, then, we are only concerned with the righthand side of the game, since this is where all the action takes place.

All we have to do to solve this game is to calculate Olga's beliefs about Igor's actions given her choices. She has to calculate Igor's expected utility for believing against his expected utility for not believing, given her two truth strategies.

$$EU(T|B) = 10p + 0(1 - p) = 10p$$
$$EU(T|NB) = 2p + 10(1 - p) = 2p + 10 - 10p = 10 - 8p$$

When $p > 0.55$, $EU(T|B) > EU(T|NB)$ and Igor should believe; when $p < 0.55$, $EU(T|NB) > EU(T|B)$ and he should not believe; and when $p = 0.55$, he should be indifferent between B and NB.

It is unclear whether Olga and Igor will get together, but the fact that Igor is infatuated with Olga makes this a win-win game for him. It does not bother him that much if Olga is a liar. Furthermore, the sequential nature of the game and the fact that Olga has to move first tips her hand in the conflicting interests game, since she has to live with the fact that if she tells Igor the truth that she hates him, he will just shrug it off as if no words had been spoken. So Olga prefers S1, in which she is nice and likes Igor. But the outcome of the game depends on Igor's beliefs about p and $(1 - p)$, since his beliefs are such that he should believe her most (though not all) of the time. Consequently, there is an equilibrium to this game when Igor believes Olga, an equilibrium when Igor does not believe Olga, and an equilibrium when Igor and Olga are indifferent among choices. All of these equilibria can be supported by varying values of p. One of the equilibria specifies that Olga will be altruistic and truthfully tell Igor that she likes him. Another specifies that Olga will be an egoist and tell Igor that she hates him and he will not believe her.

If played in real life, this game would be peppered with emotions, since it involves revealing intimate details of a person's life. This game would also present a moral dilemma concerning the merits of lying, which is also an emotional decision. This situation would be especially emotional, since even the participants themselves do not understand the emotions that are involved. Behavioral game theory is exciting simply because emotions are difficult to control and measure, so innovative methods have to be developed to understand their role in game theory models.

F. SENDER-RECEIVER FRAMEWORK LYING EXPERIMENT

In this section, I will examine a signaling experiment that involves lying. Gneezy (2005) was interested in how lies can be motivated by changes in wealth. That is, are people concerned with how much they gain from a lie or are they concerned with the losses that result from a lie? To test this question, he designed an experiment using the sender-receiver framework, in which the sender's payoff increases when he lies. In the first stage of the experiment, the sender and receiver's payoffs are revealed only to the sender. After receiving the payoff information, the sender is able to send one of two messages to the receiver:

Message A: Option A will earn you more money than option B.
Message B: Option B will earn you more money than option A.

In this case, if a message is a lie and the receiver believes it, then the sender's payoff increases and the receiver's payoff decreases. After the receiver sees the message, she has to decide whether to select Option A or Option B. Given this structure, Gneezy argues that it is possible to sustain an equilibrium when the receiver always follows the advice of the sender. To test the robustness of this prediction, Gneezy first conducted an experiment in which he assigned 50 subjects to the role of senders and asked them to estimate how many receivers would believe the message they sent. Subjects would be paid based on the accuracy of their predictions. The senders predicted that 82 percent of receivers would believe their message, and 78 percent of receivers in the experiment followed the senders' recommendation. Hence, most receivers believed the senders' message. This result confirms the evidence of a "truth bias," which is a psychological phenomenon in which a person hears a statement and without any supporting evidence believes it to be the truth (McCornack and Park 1986). Since the senders believe that the receivers are gullible, if they maximize their expected utility, then they have an incentive to lie.

To further test this hypothesis, three treatments were conducted that varied the cost of lying to the receiver, as outlined in Table 10.7. Notice that the sender always has an incentive to lie, since Option B is the lie and the sender's payoffs in Option B are all greater than his payoffs in Option A.[7]

TABLE 10.7 GNEEZY'S SENDER-RECEIVER GAME

		Payoff	
Treatment	Option	Sender	Receiver
1	A	5	6
	B	6	5
2	A	5	15
	B	6	5
3	A	5	15
	B	15	5

In Treatment 1, the sender's gain from lying is one dollar and the receiver's loss is also one dollar. In this treatment, 36 percent of subjects lied. In Treatment 2, the sender's gain from lying is still one dollar, but the receiver's loss has increased to 10 dollars. In this case, only 17 percent of subjects lied. In Treatment 3, the sender increases his payoff from lying to 10 dollars; in this case, 52 percent of senders lied. These results show that when the cost of a lie increases, subjects lie less often. However, when the relative profit of a lie greatly increases, then subjects lie more often.

In an additional test. Gneezy (2005) also had subjects respond to a hypothetical survey question. He gave subjects a scenario in which a car salesperson was about to close a deal on a car that was worth twelve hundred dollars. The seller knew that there was a defective oil-pump that would reduce the value of the car by 250 dollars, but the defect would only be recognized on a hot day, and the deal was being negotiated during the winter. The question posed to subjects was: Should the car salesperson inform the potential buyer about the defect? Gneezy then provided a follow-up question that substituted one thousand dollars for 250 dollars. The experiment concerned whether subjects believed the first lie was fairer than the second lie in terms of wealth lost. He found that as the punishment to the buyer increased ($250 to $1,000), subjects felt that the lie was less fair.

Gneezy's results show that subjects do not always maximize their expected utility from lying, and that subjects have an aversion to lying. People weigh their gains from a lie, as well as the loss of the person to whom they are lying. He notes (2005, 319):

> People are sensitive to their gain when deciding to lie. Second, people care not only how much they gain from a lie, but also how much the other side loses. This unselfish motive diminishes with the size of the gains to the decision maker herself.

G. PERSUASION EXPERIMENT

Lupia and McCubbins (1998) use a similar framework to examine the effect of persuasion, or the ability of one player (the sender) to influence, change, or reinforce the beliefs of another player (the receiver). Understanding the impact of persuasion in games that allow for communication is important, because this concept will help us to understand the effect that communication has on players' behavior, or, more specifically, on choice behavior. Lupia and McCubbins argue that perceived common interests and whether the receiver perceives the sender as being trustworthy and knowledgeable are necessary prerequisites for persuasion. That is, a person is more likely to believe someone whose interests are aligned with his or her own interests (as in Table 10.5). Hence, both players prefer to coordinate actions. This is in contrast to a situation of conflicting interests (as in Table 10.6), in which one player is trying to deceive another player. In addition, the sender is more likely to be persuasive when she is perceived to be knowledgeable in the sense that the sender knows something that the receiver does not. Over time, the receiver can develop a trusting or skeptical relationship with the sender that will make the receiver more or less likely to be persuaded. In short, a person is more likely to be persuaded by another person when the two share a common interest and when the person doing the persuading is perceived to be knowledgeable and trustworthy. However, since this belief about the sender's knowledge is perceived and not based on actual beliefs, a sender can persuade even when there is no common interest and the sender knows nothing.

The model that Lupia and McCubbins present is a little more complex than what I am presenting here, since it involves an agent who also makes choices, but for simplicity and clarity, I will just detail the game between the sender and the receiver. In this experiment, a sender observes or does not observe the outcome of a coin flip. Observation of the coin flip models the sender's knowledge, so that if the coin toss is observed the sender is knowledgeable, and if it is not observed she is not knowledgeable. The receiver's knowledge of the sender is represented by cases in which he knows the sender has observed or not observed the coin flip. After these conditions are satisfied, the sender communicates (or makes a prediction of information about the coin flip) to the receiver. Within this framework, different payoff schemes allow for differences in common and conflicting interests. For instance, in the case of common interest, the sender has an incentive to send a correct prediction, since this action results in a higher payoff for both players. In a case of conflicting interest, a sender has an incentive to be deceitful, since her payoff is inversely related to the receiver's payoff (i.e., if she can be deceitful and get away with it, then her payoff increases, but the payoff of the receiver decreases since he was deceived).

Lupia and McCubbins also introduce a penalty for lying. In this case, if the sender knowingly misrepresents the outcome of the coin toss and the receiver believes her, then the sender is assessed a "lying fine" and her payoff is reduced accordingly. The model varies the cost of lying from a low to a high cost (10 cents to one dollar), with the expectation that a high lying cost will induce persuasion, since lying will be deterred. A verification condition is also provided that introduces uncertainty on the part of the receiver about whether the true state is common or conflicting interest. The idea is that the sender's knowledge about the receiver's knowledge affects the sender's actions—if the sender knows that the receiver does not know the true state, then he is more likely to lie than if he knows that the receiver knows the state.

The results of the experiment measured persuasion as a string of consistent matches, or cases in which the receiver and sender's actions were aligned. Lupia and McCubbins found that when the sender and receiver had conflicting interests and only the sender was knowledgeable, then there were low levels of persuasion and the sender's advice only matched the receiver's actions 59 percent of the time. When the sender and receiver had common interest and the sender was knowledgeable, then there was a high level of persuasion and the sender and receiver's choices were aligned 88 percent of the time. In the low-cost lying treatment the sender's advice only matched the receiver's actions 56 percent of the time, and in the high-cost lying treatment advice and actions matched 89 percent of the time. In the verification experiments when the sender knew the receiver knew the true state, the sender's advice matched the receiver's actions 88 percent of the time.

These experiments show that communication among players in a game theory model can potentially bias results if one player is more persuasive than another player. The persuasive player can convince another player to make choices contrary to his self-interest by persuading this player to rely on bad information when making a choice. These experiments show that certain institutions allow for increased (or decreased) persuasive behavior. Lupia and McCubbins conclude (1998, 227):

> Voters, legislators, and jurors delegate to others and tend to lack information about the consequences of their actions. In many cases, these actors have opportunities to obtain knowledge from the endorsements or testimony of others. The ability of voters, legislators, and jurors to make reasoned choices and delegate successfully depends on whether their opportunities to gain knowledge actually produce knowledge. Were people able to discern the interests and expertise of others, they could make choices about whom to believe that would generate the knowledge they need. In many cases, however, people lack this ability because they do not know one another well.

H. SUMMARY

In this chapter, I have attempted to show how various imperfect and incomplete information structures can be represented in the extensive form. Obviously, the examples provided here are not comprehensive, but they do illustrate a few of the variants. I have also tried to show how these games are solved using sequential equilibrium, which is essentially the same method we used to solve mixed strategies in Chapter 7.

I introduced a subset of incomplete information extensive form games called signaling games. These games model situations in which a sender has private information about his type and conveys or does not convey information to a receiver, who must take an action based on the signal she receives. I also presented an experiment by Gneezy (2005) that illustrates the sender-receiver framework in terms of lying. The results showed that players do not always maximize their expected utility when lying is involved. Next, I presented a persuasion experiment by Lupia and McCubbins (1998) that illustrates the effect that persuasion can have on beliefs and actions.

NOTES

1. A small body of literature exists that examines games with machines that assume imperfect recall. A machine plays strategies in infinite repeated prisoner's dilemma games but only has a certain amount of memory, so that after it runs out, a player forgets what move she has made (see Osborne and Rubinstein 1994).
2. In the next chapter we will consider a third requirement that demands that beliefs at each node be updated using Bayes' Theorem. This would establish a Bayesian equilibrium.
3. This game is referred to as the little horsey game because Selten (1975) created a game similar to this in which a second subgame was placed alongside the first, with a line connecting the top of each. Hence, the diagram looked like a horse and was referred to as Selten's horse. Removing one of the subgames rendered it a "little horsey." See Problem 10.5 for an example of the horsey game.
4. Signaling games have been used in a wide range of theoretical applications. To cite a few applications, Matthews (1989) used this framework to study the president's threat of a veto on budgets proposed by Congress; Stein (1989) used it to study how announcements by the

Federal Reserve Board influence the stock market; Gilligan and Krehbiel (1987) used it to study the relationship between Congress and committees; Rogers (2001) used it to study the interaction between legislative and judicial branches of government; Austen-Smith (1990) used it to study strategic information transmission in congressional debates; and Walsh (2007) used it to study international conflict. There is also a substantive literature in evolutionary biology that uses signaling games to model animal behavior (Maynard Smith and Harper 2003).

5. There is another type of game similar to the signaling game called a screening game, in which the uninformed player moves first.

6. This labeling can be confusing in terms of truth having a dual meaning of both liking Igor and hating Igor. But remember that in each state Olga feels differently toward Igor, so that the truth about how she feels about him also varies with these feelings. Think of it like this: Let the left strategies on the top and bottom represent the message S1 and let the strategies on the right represent the message S2. If the game dictates that Olga tell Igor her state, then she lies when the state is S1 and she gives Igor the message S2 and she tells the truth when the state is S1 and she gives the message S1.

7. In his original table, Gneezy used the labels "Player 1" and "Player 2," but I have relabeled these to be "Sender" and "Receiver" for clarity.

CHAPTER 11

BAYESIAN LEARNING

A. *THE PEOPLE OF THE STATE OF CALIFORNIA V. COLLINS* (1968[1])

In June 1964 in San Pedro, California, a white woman with blond hair tied into a ponytail was seen running away from a crime scene with a stolen purse before jumping into a yellow car driven by an African American man with a beard and mustache. The couple were eventually arrested, charged with the crime, and brought to trial. During the trial, witnesses were unable to make a positive identification of the two suspects and could only identify certain characteristics of the crime: (1) African American with beard, (2) man with mustache, (3) girl with blond hair, (4) girl with ponytail, (5) interracial couple in car, (6) partially yellow car.

At the trial, the prosecutor called a statistician as an expert witness to calculate the probability that the Collinses committed the crime based on the six "clues." The statistician testified that the product rule of independent events could be used to calculate this probability. If two events are assumed to be independent, then the probability of both those events occurring is calculated by multiplying the probability of one event by the probability of the other event. This number is called a conditional probability. Statistical independence means that in order for Event A and Event B to be conditional probabilities allowing for interpretation, they must be two random variables (i.e., they cannot be related to each other). For example, assume that you want to construct a measure for a person's intelligence and you use two variables: IQ scores and level of education. Can we use conditional probability to create a measure for smartness with these two variables? No, because although IQ score and education both impact smartness, it is difficult to assume that both affect smartness *independently*, since the two are related to each other. Because these variables do not satisfy the requirement of independence, the properties of conditional probabilities do not apply. This example is simple, but in certain cases it is hard to determine if two variables are independent because of the interaction of other variables—for instance, income impacts both IQ and education level, which impact smartness.

During the Collins trial, to calculate the probability that the couple committed the crime, the prosecutor assumed that the probability of an African American was one in 10, a man with a mustache was one in four, a white woman with blond hair was one in three, a woman with a ponytail was one in ten, an interracial couple in a car was one in one thousand, and a yellow car was one in 10. The statistician used the product rule on these probabilities to calculate the conditional probability that a random couple would fit all the clues and produced a probability of one in 12 million. Although the Collinses had a sound alibi, they were convicted based on the statistical assumption of independence, which showed that there was a high probability that they had committed the crime. The conviction was appealed and overturned based in part on the assumption of independence.[2] In the appeal judgment, the justices ruled:

> There was another glaring defect in the prosecution's technique, namely an inadequate proof of the statistical independence of the six factors. No proof was presented that the characteristics selected were mutually independent, even though the witness [the statistician] himself acknowledged that such condition was essential to the proper application of the "product rule" or "multiplication rule." ... To the extent that the traits or characteristics were not mutually independent (e.g., Negroes with beards and men with mustaches obviously represent overlapping categories), the "product rule" would inevitably yield a wholly erroneous and exaggerated result even if all of the individual components had been determined with precision.

This case is used in law schools to illustrate why statistical evidence not connected with physical evidence is inadmissable in a court of law.

B. CONDITIONAL PROBABILITIES

The method of calculating conditional probabilities is important for us, because we can use this approach to solve for a Bayesian Nash equilibrium. As noted previously, a conditional probability is the probability of one independent event multiplied by the probability of another or several independent events. Let us assume that the probability of Event A is $p = 0.70$, and the probability of Event B is $q = 0.30$. The conditional (or joint) probability of Event A given Event B would be pq, or 0.70×0.30. The standard notation for a simple conditional probability of Event A given Event B is $pr(A|B)$ or, for more complex conditional probabilities given several other events, $pr(A|B, C, D, E)$.

Most game theory models only deal with simple conditional probabilities. Reconsider our lying-truth game from Figure 10.15. In this game, the conditional probability that Olga lies in State 2 and Igor believes in State 2 would be $p(L_{1,2}|B_{2,2})$, or the probability that Olga (Player 1) lies in State 2 conditioned on the probability that Igor (Player 2) believes Olga in State 2. Assume that Olga lies 50 percent of the time in State 2, and that when this happens Igor believes 50 percent of the time in State 2; the conditional probability would be 0.50×0.50, or 0.25. If Olga lies 70 percent of the time and Igor believes 30 percent of the time, then the conditional probability would be 0.21.

C. CONDITIONAL PROBABILITIES AND THE BELIEFS OF VIDEO GAME CHARACTERS

Conditional probabilities can be used to build belief systems for players, since they connect different players' strategy choices and beliefs, allowing us to examine how changes in beliefs change strategy choice and thus impact game play. These conditional probabilities become a player's belief system. Most game theory models only deal with simple conditional probabilities, producing simple beliefs that are relatively easy to calculate and interpret. However, other environments can be rich and complex, requiring beliefs about more random variables.

Consider a first-shooter video game called *Unreal Tournament* that features a synthetic character *bot*, who is controlled by a human user and engages in head-to-head multiplayer deathmatches.[3] This game is known for revolutionizing game play by introducing artificial intelligence (AI) in conjunction with the customization of a *bot*, which allows the user to vary such parameters as the accuracy of *bot*'s weapon, *bot*'s awareness of his or her surroundings, and so on. This technique adds more action to game play by making the user's movement or control of *bot* more fluid.

In this game, *bot*'s life is monitored by a health level (like a gas meter in a car) that decreases when he is hurt; when this level reaches zero, *bot* is dead. Various weapon upgrades (ammunition) and health bonuses (which increase the health meter) are scattered throughout the environment. *Bot*'s AI behavior is governed in part by conditional probabilities over a set of random variables. At one time in the game, four random variables might include (Hy et al. 2004, 179):[4]

H: whether *bot*'s health level is full at time t
W: whether *bot*'s weapon is full of ammunition at t
OW: whether the opponent's weapon is full of ammunition at t
NE: whether an enemy is nearby at t

These variables specify *bot*'s behavior within the game environment, or the variables over which *bot* has beliefs. If these were the only variables in the game, we could think of *bot*'s overall behavior as a function of the conditional probability of all of these variables together.[5] Consider that the probability of an action is conditioned on the probability that *bot*'s health is at the full level (H), his weapon is full of ammunition (W), his enemy's weapon is full of ammunition (OW), and an enemy is near (NE). Now consider three actions based on the conditional probabilities of the four variables:

1. The probability of an attack is high when the probability of H is high, W is high, OW is low to high, and NE is high.
2. The probability of *bot* fleeing a battle is high when the probabilities of H and/or W are low and NE and OW are high.
3. The probability of *bot* searching for health bonuses is high when the probability of H is low, W is high, and OW and NE are low.

Dynamic behavior is created in *bot* through changes in the variables from state S_t to S_{t+1}, with the probability value associated with each variable changing as a function of how other probability values associated with other variables change. What other types of dynamic behavior can you specify based on these four variables?

This example shows how complex belief systems can be built using conditional probabilities.[6] Game theory models normally only assume beliefs over one random variable, so we will not confront this complexity here, but it is important to note that as more complex learning structures are assumed in decision-making models, richer environments result.[7]

D. BAYESIAN LEARNING

1. WHAT IS LEARNING?

Learning concerns the dynamic behavior of rational players within a model. Thus far we have been concerned with static beliefs, or beliefs defined by a simple probability distribution. When we talk about learning in a game theory (or expected utility) sense, we are referring to how players incorporate new data (or information) into their belief systems. Recall that a belief system is a player's probability assessment of the likelihood that he or she is in a particular state, and recall that these beliefs impact payoffs. Hence, any information that is relevant has to be specific to a player's payoff. Learning occurs when a player uses new information to change his or her likelihood of being in a particular state. To be relevant, the information must convey news that either increases or decreases the player's probability of being in a particular state. If the information has no impact (i.e., it neither increases nor decreases the likelihood), it is not relevant to game play.

Information can reveal data about events that occur at any time prior to a player's decision choice, which implies that these "new information events" are somewhere further up in the game tree from the player's decision node. An event could be one decision node ahead of a player or many decisions nodes ahead of a player. As a result, actions that take place prior to a player's selection of a strategy might change her probability assessment of her current situation (e.g., casting doubt on her beliefs about her current location in the game tree). This process of recalculating probability beliefs based on new information is what we refer to as learning. Think of this concept in the following terms: Assume that I specify a particular utility function, strategy choice, and beliefs for a three-player game. I derive equilibrium for the model and then decide that I want to see what will happen to the equilibrium prediction when I remove one of the players. Learning would occur if the players could figure out for themselves what to do. This would require them to:

1. Identify and interpret new information (players must determine what probabilities have been removed).
2. Recalculate existing beliefs to form updated beliefs (players must consider the removed probabilities when recalculating their current probabilities).
3. Relate the updated beliefs to current strategy choice (forming the basis for players' new strategy choices).

As human beings, we are exposed to new data every day, some of which we discard and some of which we incorporate into our belief system. The data that we incorporate into our belief system have the potential to alter the way in which we make decisions. For instance, when a traveler enters a foreign country for the first time, she is bombarded with new information, some of which she understands and some of which she does not. She will incorporate the understandable information into her beliefs when taking an action. Consequently, in our models we will consider the process through which players are exposed to new information, and if this information impacts players' beliefs, it will be incorporated into their belief systems. When players make a decision, they should have all the relevant information that impacts their beliefs, including any new information, at their disposal.

Bayesian decision-making provides a fairly easy method to update beliefs in the sense that we have just discussed. In Bayesian decision-making, players have established beliefs called priors. When new information is introduced (in the form of "the probability of some random event has just increased or decreased"), players recalculate current beliefs to form updated beliefs called posterior beliefs.

2. UPDATING BELIEFS

Since beliefs are modeled as probabilities, it is possible to update them using probability theory. As we will see a little later, the process of updating beliefs is primarily governed by Bayes' Theorem (or rule). Before I define Bayes' Theorem, I want to reemphasize what I mean by updating beliefs. A belief is updated when some new information to which a player is privy either reinforces or changes that player's beliefs. For instance, assume that Olga and Igor have finally agreed to just remain friends, and to celebrate, they decide to take a walk at midnight. While walking together, they suddenly see a flash of light streak through the dark sky, quickly descend, and then disappear into a nearby forest. Unsure of what she just witnessed, Olga exclaims to Igor: "I just saw a UFO!" This statement represents Olga's prior (initial) beliefs about whether she saw a UFO. Let us say that Olga is 60 percent sure she saw a UFO. New information such as confirming or disconfirming statements from Igor can alter Olga's prior beliefs. She can incorporate this additional information into her prior beliefs to form a new set of beliefs called her posterior beliefs. Since Igor is a friend, his observations (or new information) will be valuable, since it comes from a reliable source. Two questions are: How does Olga incorporate Igor's new information into her beliefs, and does this new information increase or decrease her belief that she saw a UFO? The answers all depend on the consequences of the different states that are involved. If Igor remarks, "I didn't see anything," then this information could reduce Olga's posterior beliefs so that she no longer believes she saw a UFO. If Igor remarks, "Yeah, I thought that was a UFO also," then this information could increase Olga's posterior beliefs so that she becomes certain of having seen a UFO.

3. CALCULATING BAYES' THEOREM[8]

To understand how the process of updating beliefs works, let us consider two separate and independent states that are depicted in the decision tree in Figure 11.1.[9,10] In this example, in State 1, Olga thinks that there is a 60 percent chance that she saw a UFO, but Igor is more certain and thinks that there is a 70 percent chance that he saw a UFO. In State 2, Olga does not really think that she saw a UFO and figures that there is only a 40 percent chance that she saw one. In this state, Igor is also uncertain and thinks that there is a 50 percent chance that he saw a UFO. Depending on whether the players are in State 1 or State 2, a response from Igor about whether he saw a UFO or not will impact Olga's beliefs about whether she saw a UFO.

In this tree, several conditional probabilities can be produced, given Olga and Igor's beliefs. Olga and Igor can both agree that they saw a UFO, one can claim to have seen a UFO and the other can disagree, or both can agree that neither saw a UFO. Recall that the conditional probability is the belief of one player multiplied by the belief of another player in a particular state. For instance, in State 1, Olga's belief about having seen a UFO is multiplied by Igor's belief about having seen a UFO, 0.60×0.70. Figure 11.2 illustrates this situation and specifies the conditional probabilities for each of the four events.

For ease of notation, let us refer to the two states as A and B and label the case in which both think that they saw a UFO as U and the case in which they both doubt they saw a UFO as D. The four conditional probabilities (Pr) that occur in both states are the probability that Olga thinks U_O and Igor thinks U_I, or $\Pr(U_O, U_I)$; the probability that Olga thinks U_O and Igor thinks D_I, or $\Pr(U_O, D_I)$; the probability that Olga thinks D_O and Igor thinks U_I, or $\Pr(D_O, U_I)$; and the probability that Olga thinks D_O and Igor thinks D_I, or $\Pr(D_O, D_I)$.

Bayes' rule is simply the conditional probability of one event in State A divided by the probability of that event plus the corresponding event in State B. The denominator, which is determined by summing across both states, ensures that the event occurs regardless of

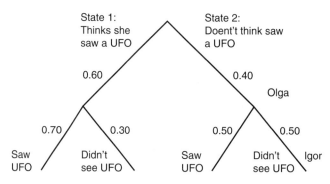

FIGURE 11.1 Decision tree

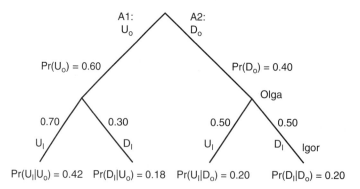

FIGURE 11.2 Conditional probabilities

whether it is State A or State B. If the event is the same in State A and State B and yields the same payoff in both, then no new information is granted and the priors will not change. In this situation, a player is in a complete information environment. An incomplete information environment implies that greater payoff differences in the two states allow for greater consequences of actions (i.e., higher or lower payoffs).

While there are many technical ways to formulate Bayes' rule, I will present an informal one that is defined in terms of our discussion about states and events.[11]

Bayes' rule is the probability of:

$$\frac{\text{Event in State A}}{\text{Event in State A} + \text{Same Event in State B}}$$

Let us take one event in State A: both Olga and Igor think that they saw a UFO, producing a conditional probability of $Pr(U_{OA}, U_{IA})$. What is this same event in State B? In this case, Olga does not see a UFO, but Igor does (D_{OB}, U_{IB}). Note that Igor takes the same action in both states. The other possible event in State A would be for Olga to see a UFO but for Igor to not see it (U_{OA}, D_{IA}). In the same event in State B, neither Olga nor Igor would see a UFO (D_{OB}, D_{IB}). For Case 1 (both Olga and Igor see a UFO), Bayes' rule would be:

$$\frac{Pr(U_O, U_I)}{Pr(U_O, U_I) + Pr(D_O, U_I)}$$

For Case 2 (Olga sees a UFO and Igor does not), Bayes' rule would be:

$$\frac{Pr(U_O, D_I)}{Pr(U_O, D_I) + Pr(D_O, D_I)}$$

For Case 3 (Olga does not see a UFO but Igor does), Bayes' rule would be:

$$\frac{Pr(D_O, U_I)}{Pr(D_O, U_I) + Pr(U_O, U_I)}$$

And for Case 4 (neither Olga nor Igor sees a UFO), Bayes' rule would be:

$$\frac{Pr(D_O, D_I)}{Pr(D_O, D_I) + Pr(U_O, D_I)}$$

In order to solve these equations, we need the following probabilities:

$Pr(U_O) = 0.60$
$Pr(D_O) = 0.40$
$Pr(U_O, U_I) = 0.70$
$Pr(U_O, D_I) = 0.30$
$Pr(D_O, U_I) = 0.50$
$Pr(D_O, D_I) = 0.50$

Now we can simply plug these probabilities into the equations above and get the posterior beliefs:

$Pr(U_{OB}, U_{IA}) = 0.70 \times 0.60 / (0.70 \times 0.60) + (0.50 \times 0.40) = 0.6774$
$Pr(U_{OA}, D_{IA}) = 0.50 \times 0.40 / (0.70 \times 0.60) + (0.50 \times 0.40) = 0.322$
$Pr(D_{OB}, U_{IB}) = 0.30 \times 0.60/ (0.30 \times 0.60) + (0.5 \times 0.4) = 0.47$
$Pr(D_{OB}, D_{IB}) = 0.5 \times 0.4 / (0.5 \times 0.4) + (0.30 \times 0.60) = 0.53$

According to the updated priors—the posterior beliefs—in State A, when Olga thinks she saw a UFO and this belief is reinforced by Igor's beliefs that he saw a UFO, her prior beliefs increase from 60 percent to 68 percent, leaving her more certain of having seen a UFO. However, in State B, when both she and Igor doubt that they saw a UFO, Olga's posteriors hover around 50 percent, and she does not believe that she saw a UFO.[12] Bayes' rule provides a method to model how players incorporate other players' beliefs (i.e., learn new information from an external source). This new belief, which incorporates the new information, is a posterior belief, or the type of belief that is used when a decision is made.

I should note that Bayesian updating is just one learning model, and I will discuss alternative learning models near the end of this chapter. The question of how people update their beliefs when new information is presented to them is a subject of great debate. People learn in many different ways, because the process of absorbing new information into a belief system depends, in a large part, on the type (relevance) of information that is being absorbed, the manner in which the information is presented, and the source of the information. As a learning model, Bayesian decision-making provides a good baseline or framework in which to study other types of learning. In many respects, Bayesian decision-making is a good learning model, since it does mimic ways in which actual people process actual information. The problem is whether people update their beliefs precisely, or at all. Furthermore, the probability numbers produced by Bayes' rule can often be hard to interpret. For example, what if a player's prior belief about an event increases from 49 percent to a posterior of 51 percent? Is the player uncertain at 49 percent and certain at 51 percent, or is he still just uncertain?

E. PERFECT BAYESIAN EQUILIBRIUM

1. WEAK CONSISTENCY OF BELIEFS

Now that we have learned that beliefs in a game are modeled as conditional probabilities and that belief updating is dictated by Bayes' rule, I will introduce the concept of perfect Bayesian equilibrium (Kreps and Wilson 1982; Fudenberg and Tirole 1991). This concept is simply a generalization of a sequential equilibrium and, in most instances, is the same as a sequential equilibrium. The difference between a sequential equilibrium and a Bayesian equilibrium is how beliefs are modeled. In a sequential equilibrium, beliefs are simple, not conditional, probabilities. In contrast, to establish Bayesian equilibrium, we must assume a weak consistency condition or "weak consistency of beliefs." Loosely, this phrase means that for every information set on the equilibrium path that is reached with positive probability, beliefs are defined by Bayes' rule. That is, a player's beliefs follow Bayes' rule and are optimal given the choices and beliefs presented by another player. We can add this weak consistency requirement to the list of requirements for sequential equilibrium outlined in Chapter 10.

This requirement is used when there are mixed strategies or the probability of nature is at the initial information nodes, in which case we calculate expected payoffs using beliefs. Consider the game in Figure 11.3.

In this game, the initial node includes a mixed strategy for Player 1 that consists of playing M1 with probability p and playing R1 with probability q (with p and q summing to 1). Hence, one restriction is that if the equilibrium path includes either M1 or R1, then consistency requires the values of p and q to be calculated using Bayes' rule, or else Player 2 must assign the belief $p/(p + q)$ to M1 and $q/(p + q)$ to R1. This assumes that if Player 2 knows that Player 1 is going to play M1 with certainty or with a probability of 1, then Player 2 must assign probability p = 1 for M1 and q = 0 for R1; similarly, if he believes that Player 1 will play R1 with certainty, then he must assign q = 1 and p = 0.[13] If the equilibrium path includes L1, then no restrictions are placed on p or q, since there is no information set for Player 2 at this decision node.

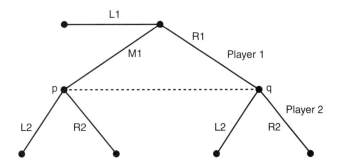

FIGURE 11.3 Consistency requirement

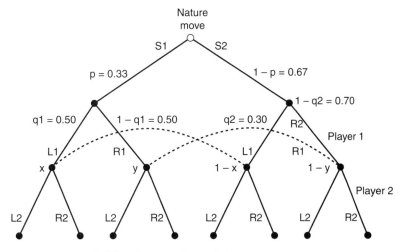

FIGURE 11.4 Beliefs based on Bayesian updating

To illustrate this requirement, consider the game in Figure 11.4. In this game, there are two states of nature, for which State 1 occurs with probability p = 1/3 and State 2 occurs with probability (1 − p) = 2/3. Player 1 has the first move and must decide whether to select L1 or R1. Since there is no information set connecting Player 1's decision nodes, she knows which state she is in when she makes a decision. Player 2, however, faces uncertainty, because he only knows if Player 1 has selected L1 or R1 and not the state in which he is located. In State 1, Player 1 selects L1 with probability q1 and R1 with probability (1 − q1), and in State 2, she selects L1 with probability q2 and R1 with probability (1 − q2). For this example, assume that q1 = 0.5 and q2 = 0.3. Now, when L1 is played, Player 2 has probability x of being in State 1 and probability (1 − x) of being in State 2, and when R1 is played, he has probability y of being in State 1 and probability (1 − y) of being in State 2.

Note that there are no values for Player 2's beliefs (x or y) in the figure. That is because Player 2's beliefs are restricted by the consistency requirement to be based on Bayes' rule. This requirement forces Player 2's beliefs to be based on the sequence of events and the resulting conditional probabilities at the node at which he is making a decision (at x and y). We can easily solve for x and y in this game. To solve for x,

$$\frac{\Pr(p, q1)}{\Pr(p, q1) + \Pr(1 - p, q2)}$$

To solve for y,

$$\frac{\Pr(p, 1 - q1)}{\Pr(p, 1 - q1) + \Pr(1 - p, 1 - q2)}$$

Plugging in the values like we did before yields

X = 0.166/0.496, or 0.33
Y = 0.166/0.632, or 0.262

Now we can substitute these values into our game, as in Figure 11.4, and solve the game in the same way that we would solve for the sequential equilibrium. It is important to note once again that in games of imperfect and incomplete information, different values of beliefs can sustain different strategies in equilibrium.

2. SOLVING FOR A PERFECT BAYESIAN EQUILIBRIUM[14]

There is no standard way or algorithm to solve for a perfect Bayesian equilibrium (PBE). Recall the discussion of separating and pooling equilibrium from the last chapter, which indicates that one way to solve for a PBE would be to determine if one of these equilibria exist. In this approach, first see if there are any Nash equilibria by looking at the normal form game. This knowledge allows us to determine if these Nash equilibria can be rationalized in the sense that strategy choices are sequentially rational given a set of beliefs.

If we examine the normal form of the game depicted in Figure 11.5, we can see that there is a Nash equilibrium in State 1 with strategies (L1, L2) and payoffs (1, 2), and a Nash equilibrium in State 2 with strategies (R1, R2) and payoffs (2, 1). To find a PBE, we must

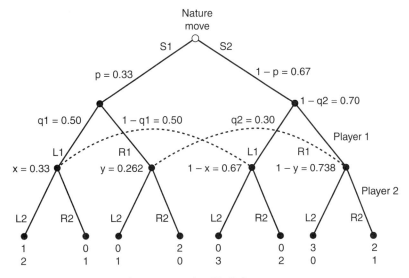

FIGURE 11.5   Game with Bayesian updated beliefs

determine whether there is a set of beliefs that can sustain these equilibria. We must first consider Player 1's beliefs about Player 2 given her choices and his choices and then calculate the expected utility for each action.

When Player 1 plays L1, Player 2's expected utility from playing L2 is

$$EU_2(L2) = 0.33(2) + 0.67(3) = 2.67$$

When Player 1 plays L1, Player 2's expected utility from playing R2 is

$$EU_2(R2) = 0.33(1) + 0.67(2) = 1.67$$

When Player 1 plays R1, Player 2's expected utility from playing L2 is

$$EU_2(L2) = 0.262(1) + 0.734(0) = 0.262$$

When Player 1 plays R1, Player 2's expected utility from playing R2 is

$$EU_2(R2) = 0.262(0) + 0.734(1) = 0.734$$

Now, Player 1 knows that if she selects L1, then Player 2 will select L2, and if she selects R1, then Player 2 will select R2. Player 2 knows that if he sees L1, he is in State 1 and should select L2; if he sees R1, he knows he is in State 2 and should select R2. A PBE for this game can be sustained with a system of beliefs such that L1 and L2 are selected in State 1 when x = 1. Another PBE can be sustained with a system of beliefs such that R1 and R2 are selected in State 2 when y = 0. In these two cases, all strategies are consistent with beliefs.

This simple example can be solved using the tools we have learned in this book. However, there is no standard algorithm that can be used to find a PBE. In some games, the only way to solve for a PBE is to try different levels of beliefs to see if a strategy profile is consistent with a certain set of beliefs.

3. REFINEMENTS TO PERFECT BAYESIAN EQUILIBRIUM

This chapter aims to introduce Bayesian learning and illustrate how this type of learning model is used to derive equilibrium in game theory models. There are a host of other issues on which I have not touched, with the most important being beliefs that occur off the equilibrium path. Thus far, our theory of decision-making has only been concerned with behavior that occurs on the equilibrium path. However, there has to be concern for behavior and beliefs when we are considering cases in which a player has no move, because such behavior can impact behavior that occurs on the equilibrium path. That is, in an extensive form game, some information sets might not be reached with a non-zero probability, making Bayes' rule inapplicable. Consider the game in Figure 11.6.

In this game, a PBE is reached when Player 1 selects R1, Player 2 selects R2, and p = 0. However, this game also has a Nash equilibrium, which occurs when Player 1 selects L1 and Player 2 selects L2. This Nash equilibrium is not supported by a PBE, because it requires Player 2 to select L2 when consistency requires him to select R2.

When information sets are off the equilibrium path, any beliefs can be assigned to them. Weak consistency places no restrictions on beliefs that are "off the equilibrium" decision nodes. Therefore, a fourth requirement for behavior simply states that we should use Bayes' rule whenever possible. Equilibrium refinements to PBE are concerned with modeling behavior by placing restrictions on beliefs that are off the equilibrium path. Refinement means examining multiple equilibria and, according to some set of criteria, selecting one of the equilibria as more plausible than the others. One way to fulfill this requirement is to specify a "reasonable" belief off the equilibrium path (for signaling games, see Cho and Kreps' [1987] intuitive criterion).

Restrictions are also needed to deal with questions of common knowledge. Thus far, we have only been concerned with two players. But what happens if we are modeling a three-player game? Should two players share a common belief about a third player, or should each player have a different belief about this third player? What happens if the game has more than three players? For more information about these types of refinements, see Fudenberg and Tirole (1991), and van Damme (2002).

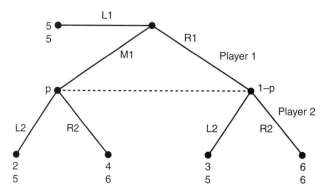

FIGURE 11.6 Equilibrium (L1, L2) not supported by a PBE

F. INFORMATION CASCADE EXPERIMENTS

In this book, we have mostly restricted ourselves to two-player games; however, interesting games can be modeled using many players. In this section I will discuss sequential move games of perfect but incomplete information with many players. In a sequential move game, *n* players are in some sort of queue. One player goes first and takes some action, then another player takes some action, then another player takes some action, and so on. This type of game creates an environment in which some players are more uncertain than others, as opposed to a simultaneous move game, in which all of the players move and take actions at the same time, making each player equally uncertain. The sequential structure is a peculiar institution that is not often formally used, but it models an environment with unequal uncertainty that is commonly understood—that is, an environment in which some people are more uncertain than others when a joint decision is being made. The important question, then, is: How does this sequential information structure bias outcomes? Some examples of sequential voting in the real world are ratification of the U.S. Constitution, U.S. presidential primary elections, and European Union elections. Other institutional voting includes conference voting in Supreme Court opinion decisions (in which the vote on the final opinion is conducted in order of seniority) and roll-call voting (in which members of legislative bodies cast votes at different times).

Sequential move games can vary players' level of uncertainty based on their position in the queue. These games model situations of perfect information (in which players know their own and others' position in the queue) and incomplete information (players do not know other players' preferences). Assume that three players have to use majority rule vote to decide a favorite color between blue and green, and that they have to make their choice sequentially. Each player is randomly given a blue or a green card that represents his or her most preferred color, or his or her private information. Players know their preferred color but not the preferred color of other players. Each player starts off with 50/50 beliefs about the preferred color of other players. Player 1 goes first and reveals her preferred color by voting for blue or green. Her vote is based solely on her private information, since her uncertainty about others' preferred colors remains the same. Let us assume that she selects and votes for green. Player 2 now votes and is privy to two pieces of information before he does, his color preference (let us say blue) and Player 1's preference. He is only uncertain about Player 3's preferred color. Let us say that Player 2 votes for blue. Player 3, who goes last, is no longer uncertain at all, since she knows the color preferences of both Player 1 (green) and Player 2 (blue), as well as her own preference. Player 3 therefore knows all there is to know. Given this sequence of colors, Player 3 can select her most preferred color as the favorite color.

This type of information structure can cause an information cascade, in which some players free-ride off the information of other players or cause some players to disregard their private information (e.g., their preference in color) in favor of some sort of public information (e.g., vote totals). Anderson and Holt (1997, 847) explain that an information cascade "occurs when initial decisions coincide in a way that it is optimal for each of the subsequent individuals to ignore his or her private signals and follow the established patterns." However, cascades are fragile in the sense that they can be manipulated to yield biased outcomes—that is, outcomes that would yield higher payoffs (both individually and collectively) if the players abided by their own private information instead of discarding their private information in favor of some sort of public signal (see Banerjee 1992; Bikhchandani, Hirshleifer, and Welch 1992; Allsopp and Hey 2000; and Hung and Plott 2001).

An experiment by Anderson and Holt (1997) shows how this type of institution can induce information cascades. In this experiment there are two urns labeled A and B, and each urn has a different distribution of marbles. One of the two urns is randomly selected without the player's knowledge, and the player must predict which of the two urns has been selected. Urn A has two white marbles and one black marble, and Urn B has two black marbles and one white marble. The posterior probability for a white ball is 2/3 in Urn A and 1/3 in Urn B. Three people are asked to select one marble (which is then replaced) and conceal its color from other players. Afterwards, each player makes a public prediction about the urn from which he or she thinks the marble came. As in the previous example, our three people take action in a sequential manner, which allows them to observe actions of players who are ahead of them in the queue. Notice how a cascade can start from the following action. Assume that Player 1 selects a white marble and predicts that it comes from Urn A (the correct prediction, given beliefs). Player 2 then selects a black marble and, instead of predicting Urn B like his private information suggests he should do (since it is more likely to be from Urn B), predicts Urn A, thinking that Player 1 could be correct. This sets up Player 3, who selects a black marble and, upon learning that Player 1 and Player 2 have predicted A, predicts A (despite her own private information indicating that it is probably from Urn B) with the herd. Given the three picks (two black and one white), the correct prediction would have probably been Urn B. In this instance, Player 2 started a cascade by causing an imbalance. Anderson and Holt observe that "an information cascade is possible if an imbalance of previous inferred signals cause a person's optimal decision to be inconsistent with his or her private signal" (1997, 851). Hence, an imbalance occurs when the posterior beliefs are 1/2. The difference in the number of white marbles and black marbles determines the differences in posterior beliefs. The posteriors for the three draws of the marbles are listed in Table 11.1.

To understand this table, first focus on the first column of fractions. The labels of zero associated with the first fraction 1/2 mean that with no draws of a marble, the posterior probability of the marble coming from Urn A or Urn B is 50/50. Going down the column, notice that if one white marble is drawn and revealed, then a player's posterior probability that the urn is A will be 2/3; if two white marbles are drawn, then the posterior probability of it being Urn A will be 4/5; and if three white marbles are revealed, then the posterior

TABLE 11.1 POSTERIOR PROBABILITIES

		Black marbles			
		0	1	2	3
White marbles	0	1/2	1/3	1/5	1/9
	1	2/3	1/2	1/3	1/5
	2	4/5	2/3	1/2	1/3
	3	8/9	4/5	2/3	1/2

Source: Anderson and Holt 1997.

probability of it being Urn A will be 8/9. Conversely, if three black marbles are drawn, the posterior belief of it being Urn A will be 1/9. The interesting cases that the authors identify as imbalanced states are those cases where the probabilities are at 1/2 on the diagonal line of the table. In these cases, a player confronts a tied state (1/1, 2/2, 3/3) and can subsequently break the tie and start a cascade in his favor from scratch.

Experiments were conducted in which subjects followed the procedures described previously. Subjects were paid two dollars for correct predictions and nothing for incorrect predictions. Subjects played 15 rounds of the experiment. At the start of the experiment, a monitor threw a dice to determine which urn would be selected, with the outcome hidden from the subjects' view so they did not know the draw. The experiment used six subjects who were randomly selected for the queue in each period. In each period, subjects picked their private draw of colored marbles and made a public announcement about the urn from which they thought the marble came. All subjects recorded the predictions of all other subjects depending on where they were located in the queue. Hence, Player 1 had only her private draw to make a decision, Player 2 had his private draw and Player 1's prediction, and so on. At the end of the period, the correct urn was announced and the next period started.

The results found cascade behavior in 41 of 56 periods. Anderson and Holt (1997, 852) give an example of a cascade that begins in period five when Urn B was selected:

Round 1: Player 1 drew a white marble and predicted A
Round 2: Player 2 drew a black marble and predicted B
Round 3: Player 3 drew a black marble and predicted B
Round 4: Player 4 drew a black marble and predicted B
Round 5: Player 5 drew a white marble and predicted B
Round 6: Player 6 drew a white marble and predicted B

Notice that Players 5 and 6 predicted against their private information. These predictions were caused by the third and fourth rounds, in which players predicted correctly and created the imbalance that resulted in the cascade. These results show that the sequence in which subjects receive information can impact their behavior. As well, these experiments show that herd behavior can lead to suboptimal or optimal outcomes. In this type of decision environment, the order in which information is presented may motivate some players to ignore their private information and rely on a public signal. This behavior could be optimal if the information content of prior predictions exceeded the content of a player's own signal. We can imagine that if costs were associated with acquiring private information, there could be more of a reliance on public (costless) information, which might or might not benefit a player.

Finally, these experiments showed that some subjects conformed to Bayesian decision-making about two-thirds of the time, meaning that about one-third did not. One question that this experiment raises is: What type of decision-making heuristic are these other subjects following if it is not Bayesian? The authors provide alternative explanations (e.g., a counting heuristic) but these results are unclear. Çelen and Kariv (2004) conducted an information cascade experiment that allowed for imperfect and incomplete information and found that herd behavior occurred less frequently, indicating that knowledge about the order of the sequence matters in terms of herd behavior.

G. ALTERNATIVE LEARNING MODELS

Behavioral game theorists are interested in questions of how equilibrium arises in a game in the first place. Bayesian updating is based on assumptions of strict rationality. Other learning models lessen this restriction (called bounded rationality), which provides different answers to this question. Experiments are a good way to sort through the various alternative learning models. Camerer notes (2003, 265):

> Experimental data are a good way to test models of learning because control over payoffs and information means we can be sure what subjects know (and know others know, and so on), what they expect to earn from different strategies, what they have experienced in the past, and so forth. Since most models make detailed predictions that require this information, laboratory control is indispensable for sorting out candidate models.

While I do not have space to fully discuss all the alternative learning models, I will briefly list several to illustrate the diversity of approaches (for an expanded explanation, see Camerer 2003; and Camerer, Ho, and Chong 2002).

1. In evolutionary learning, players are not searching for equilibrium strategies, but rather are "born" in equilibrium and innately know the equilibrium strategies. In this case, players do not learn (since they already know everything there is to learn); rather, they survive.
2. Automatic imitation (or reinforcement) learning occurs when players mimic actions that have been successful in the past.[15] In this theory, people learn by mimicking the behavior of others. Behavior is not a result of thinking but is based solely on the actions of others. Hence, actions are not voluntary but are automatically generated. One concern with this model is that some actions can be taken that have not occurred in the past, so actions can be taken that are not mimicry.
3. In belief learning, players build a model of their opponent to learn to respond to his actions. There are two extremes in belief learning:
   a. In weighted fictitious play, a model of the other player's behavior is built based on past history, where weight may be given to more current actions and the past discounted.
   b. In Cournot best-response dynamics, a model of an opponent is built by assuming that he will just repeat his last action.
4. In imitation learning, players observe what other successful players are doing and imitate them. This learning model has problems explaining innovation and players who consistently beat other players.
5. In rule learning, players learn a set of plausible limiting decision rules that allow them to choose strategies. Varying different rules allows learning to be observed.
6. Experience-weighted attraction integrates rule learning and belief learning with two primary parameters: the first adjusts between cumulative beliefs and weighted beliefs and the second mixes between reinforcement learning and beliefs.
7. Cultural learning refers to the fact that people learn based on the social norms that are prevalent within their environment.

H. EQUILIBRIUM AND LEARNING

In this final section, I want to briefly revisit the notion of equilibrium and reconsider what this concept *really* means. In a physical sense, an equilibrium occurs when there is a balance between opposing physical forces (like a polarized magnet), a state in which all influences cancel each other out. In terms of game theory, an equilibrium is a state in which payoffs are stable, actions are stable, strategies are stable, and beliefs are stable. All of these elements are "stuck," "frozen," or "unmovable" as in the physical sense, but only at one point in time, since a Nash equilibrium is a snapshot of events at one particular moment. Future game theory models will need to address how this snapshot equilibrium generalizes to behavior over time when players gain new information. As long as our players are engaged in a battle and strategies are deployed, if new information is being processed, then strategies will forever refine and improve. The study of equilibrium behavior in learning environments is important because it allows us to build more comprehensive "axiomatic" models of human decision-making.

I. SUMMARY

This last chapter concerned Bayesian decision-making, which provides a method for players to update their beliefs about their opponent's expected behavior. This method, which relies on the standard assumption of rationality, does a good job theoretically as a method for updating beliefs. However, real human behavior does not always correspond to the behavior that is necessary for making the precise calculations used to solve for this solution. Instead, theorists have attempted to create alternative learning models that are just as robust as Bayesian updating but less restrictive than the PBE. I explained that a PBE is the same as an SE, with the exception that an SE relies on simple beliefs and a PBE relies on conditional probability beliefs. Finally, I discussed an experiment that measured a bandwagon effect caused by the sequential structure of decision-making. In this institution, players have an incentive to disregard their private information in favor of some public signal.

NOTES

1. http://www.law.berkeley.edu/faculty/sklansky/evidence/evidence/cases/Cases%20for%20TOA/People%20v.%20Collins.htm
2. The concern was also expressed that the "the jurors were unduly impressed by the mystique of the mathematical demonstration but were unable to assess its relevancy or value." Furthermore, the courts ruled that even astronomical odds of probability of guilt do not mean guilt and noted: "The prosecution's figures actually imply a likelihood of over 40 percent that the Collinses could be 'duplicated' by at least one other couple who might equally have committed the San Pedro robbery."
3. This discussion is motivated by Hy et al. (2004). *Bot*'s AI was created by programmer Steve Polge.
4. I have modified the variable names from Hy et al. (2004) to make interpretation easier.
5. Other variables include S_t (*bot*'s state at time t), S_{t+1} (*bot*'s state at time t + 1), HN (whether a noise has been heard at time t), PW (whether a weapon is close by at t), and OH (whether a health pack is close by at t).

6. As you can imagine, modeling the conditional probabilities of all actions given the large number of variables could quickly become a computational nightmare. The programmers in this game handled this problem by creating separate tables for a group of conditional probabilities and then connecting all the tables together to form a system of beliefs.

7. This is not to imply that game theory and video games are alike, only that players' beliefs in both cases are modeled in a similar fashion.

8. Named after Thomas Bayes in 1764 (see Bayes 1958).

9. Be advised that this is not a game, but only a tree illustrating the strategy combinations and beliefs at the time a strategy is chosen.

10. I could be convinced that these two events are not independent or do not satisfy the property of disjointness. But for illustrative purposes, let us assume that they do. The problem is that being almost uncertain in one state and almost certain in another state means that you can be certain and uncertain in both states, meaning that properties of probability theory do not hold.

11. See any of the game theory or statistics texts referenced in this book for a more formal definition and formula.

12. A more accurate statement might be that she is less certain of having seen a UFO.

13. This example is from Osborne (2004, 327).

14. This is also referred to as a weak perfect Bayesian equilibrium.

15. See Footnote 2 in Chapter 7 for a reference to automatic imitation.

CHAPTER PROBLEM SETS

CHAPTER 1: WHAT IS GAME THEORY?

1.1. Let us start thinking in terms of game theory models and turning everyday situations into games. The first step is to start thinking about strategies and what they are. Consider the following:

1. Let us assume that you are at a coffee shop pretending that you are reading but really checking out the other people there. You notice someone else who is doing the same thing and also notice that this person is attractive and is looking at you!
   a. In this situation, what are two actions you can take?
   b. What are two actions that this other person can take? (*Hint:* These actions should be the same as your actions.)
   c. What are the possible combinations of the actions? Draw a 2 × 2 table and list each combination of actions in each cell. That is, if there are two actions 1 and 2, then the possible combinations of these actions that would be listed in the cells would be (1, 1), (1, 2), (2, 1), and (2, 2).

2. Let us assume that you are at a bar and a drunk patron stumbles and falls into you, causing you to spill your drink over yourself.
   a. In this situation, what are two actions you can take?
   b. What are two actions that this other person can take?
   c. Diagram the combination of the actions in a 2 × 2 table.

3. Finally, assume that you have an exam for which you have to study and you run into a friend who also has to study for an exam. However, neither of you has seen the other for ages, and you both want to catch up.
   a. In this situation, what are three actions you can take?
   b. What are three actions that this other person can take?
   c. List the possible combinations of these actions in a 3 × 3 table.

1.2. Consider the different types of games discussed in Section E of this chapter. Pick three of the different types of games and think of how something you have observed in your daily life fits the description of each game. For instance, an example of imperfect information would be a situation in which I have a secret from my friend that, if she knew it, would change her behavior toward me. Try to provide as full of an answer as possible for the three types of games you select.

1.3. Recall that variables are "concepts" that can be measured. Think of two concepts related to outer space and describe how they can be measured. Then think of two concepts that are underground and describe how they can be measured.

1.4. A dependent variable is something you are trying to explain and an independent variable is something you use to explain the dependent variable. For example, consider the statement "I needed some coffee to stay up." The dependent variable is what you are trying to explain—"the need to stay up"—and the independent variable is what you are using to explain the dependent variable—"coffee." Consider politics and list three concepts that could be dependent variables. Formulate an independent variable for each dependent variable and explain the relationship that you would expect to exist between the two variables.

1.5. Consider the following relationships:

- A is related to B and B is related to C
- A is related to B and B is not related to C, but A is related to C
- A is not related to B, but C is related to A and B

Think of three variables for A, B, and C for which the three relationships would hold. Explain your logic for each situation.

1.6. Let us expand our understanding of explanations. Consider the relationship between a person's geographical location and the type of music to which he listens most often.
- a. Pick two types of music and associate those types of music with a geographical region (any divisions are fine as long as they are comparable, such as urban/rural, East Coast/West Coast, South/North, suburban/urban). Explain why you think that people from these geographical areas prefer a certain type of music
- b. Consider age and music preference, and specifically the two age groups of old and young. Associate a type of music with each age group and explain why you think music preference is based on age (be sure to justify your ideas of what is old and what is young).
- c. You should now have four relationships, two for geography and two for age, for each pair (e.g., old rural, young rural, old urban, young urban). Explain the taste in music that this combined group is more likely to have.

1.7. Explain a pattern of behavior in which you like someone at one moment and, for no apparent reason, do not like him or her in the next moment, although he or she has taken no action. Is this rational behavior? What could explain this behavior?
- a. Provide an explanation for this behavior, assuming that the behavior is rational and makes sense.
- b. Provide an explanation for this behavior, assuming that it is influenced by some emotion.
- c. Provide an explanation for this behavior, assuming that the behavior is irrational (use your own definition).

1.8. Consider a city that is divided into two populations, red and blues, where reds have a majority of members. The former mayor of the city is a member of the red group, but he was publicly disgraced while in office and had to serve jail time. After serving his time, the former mayor decides to run for reelection. All blues hate the former mayor and will vote for any blue opponent. However, reds are divided among themselves. Some resent the former mayor and would vote for a blue candidate over him. Others forgive the former mayor and would support him over any blue candidate. What must happen (in terms of coalitions or proportion of reds and blues) for the former mayor to get reelected? What must happen for the blue candidate to get elected?

1.9. The president has inherent authority (a power not defined in the Constitution) to use an executive agreement. This is essentially a memo from the president that is entered into the Federal Register. It is a way for the president to pass laws or legislate without having to push a bill through Congress (possibly expending valuable resources in the process). The question is: Why would Congress allow the president authority to legislate when Article 1 of the Constitution vests this authority in Congress?

Consider the following two conclusions:

1. Executive orders offer an advantage to a president because he or she can claim credit for policies while avoiding the costs of a congressional battle.
2. Executive orders offer an advantage to Congress because they can claim credit for policies while avoiding the costs of a congressional battle.

Consider the following assumptions:
   a. Citizens perceive no difference between policies introduced and passed by the president via executive orders and those passed by Congress.
   b. Some executive orders are viewed as noncontroversial policies (affirmative action for veterans) and others as controversial policies (race-based affirmative action).
   c. The president bears the resource costs of policy implementation, whereas the political costs and benefits of that policy are shared by both the president and Congress (either equally or unequally).
   d. The president has to spend resources in proportion to Congress' resistance.

Construct a 500-word theory using the assumptions to reach the two conclusions stated previously.

1.10. Judges at the state and district levels are sometimes elected by voters.

Assume that each judge has a score on a law and order dimension. For instance, consider a scale concerning judges' opinions on the death penalty, with 0 indicating a position in favor of rehabilitating criminals, 100 indicating a position in favor of the death penalty with no rehabilitation, and 50 indicating a position in between the two. Assume that two district-level judges are competing for a state supreme court seat, and that Judge C has a score of 75 and Judge P has a score of 25. Also assume:

1. Most people in the state are opposed to the death penalty.
2. Most people in the state are not favorable toward the Second Amendment.

3. Most people in the state favor the use of marijuana for medical reasons.
4. Most people in the state favor mandatory sentences for convicted felons.
5. Most people in the state are Republicans (independent).
6. Most people in the state prefer tennis to football.
7. Most people in the state know nothing about either judge.
8. Most people in the state are old.
9. Most people in the state listen to (insert most hated/favorite song).

   a. Create a law and order dimension and place each of these issues on the dimension at a location you can justify.
   b. Assume that these assumptions represent separate elections in which one of the assumptions is the only issue of concern in the election. Which judicial candidate would win in the various elections?
   c. Assume that state supreme court judges are appointed by a partisan legislature in which the majority party has a score of 60. Who will be appointed, C or P?
   d. Assume that state supreme court judges are appointed by a bipartisan commission with a score of 50. Who will be appointed?
   e. Now assume that the commission is tied at 50 and you have the deciding vote. Who will be appointed, C or P?

WHAT ARE LABORATORY EXPERIMENTS?

2.1. Causality usually involves "directional" relationships between two variables. Consider two variables: the knowledge a person has (K) and the time he spends reading a knowledgeable book (T). In terms of direction (+, -), two variables can either increase together (+, +), two variables can decrease together (-, -), one variable can increase while the other decreases (+, -), and one variable can decrease while the other increases (-, +). Use the variables K and T to provide a literal description of their relationship in the four directional states.

2.2. Think about an experimental design for the following two situations.
1. Assume that you are in a class with 20 other students and your professor gives you an assignment to determine how people deal with stress. Each of you is paired with another class member, and you have to track each other for an entire semester. At the end of the semester you will generate a report with empirical proof about the strategies this other person uses to cope with stress. Design an experiment that will generate some empirical evidence and specifies agreed-upon observational periods.
2. Many observers say that violent video games can cause violent behavior in people. Assume that you have one friend who plays these games and exhibits violent and aggressive behavior, but you also have another friend who plays the same games and is nice and mellow. Think of an experiment that you could implement that could test the effect of video games on aggressive behavior in terms of your two friends. (To answer this question, you need to think of some common difference between the two friends that can be manipulated.)

2.3. Students have different study habits. Consider three ways in which a student can study: (with music, in silent), (with others, alone), and (at home, at library). Given these three categories, explain the efficiency of each study technique and rank them in terms of your requirement of efficiency.

2.4. Consider the following experiment. In this experiment, Player 1 has 100 dollars, of which she can give any amount to Player 2. After Player 1 makes a proposal to Player 2 concerning the allocation of the 100 dollars, then Player 2 has a final move: he can either accept the proposal and the game ends, or he can impose a penalty by which all payoffs to both players are cut by 50 percent, and then the game ends.

    a. What is the purpose (dependent variable) of this experiment?
    b. Explain how the decision affects both Player 1 and Player 2.
    c. What variables need to be controlled for in this experiment?
    d. What do you think the outcome of this experiment will be? Explain.

2.5. Let us say that you want to design an experiment to isolate outcomes involving some minority group as compared to some majority group. However, you want to ensure anonymity of choices so that subjects do not know the people against whom they are playing. So you want groups to be identified but the choices of individual subjects to be anonymous. What procedure would you implement to do this? Before you consider this question, understand that it relates to the purpose of your study. There are two ways to implement controls, depending on the purpose of the study:

    a. Does the research question just involve differences in preferences between two groups? If so, then "real minorities" may not be needed, because different preferences can be assigned in the experiment.
    b. Does the research question involve differences between groups based on some psychological attachments of the groups that can only be realized by having choices from "real minority" subjects? If this is the case, then "real minorities" will have to be recruited for the experiment.

To implement a procedure for our experiment, we have to identify a research question. Consider a simple research question such as: Do "red" people who are in the majority become more "nervous" as the number of "blue" people increases? Based on this research question, does the experiment involve the first or second case described here? The answer depends on how you interpret the research question. For this exercise, first interpret the research question in terms of the cases and then describe how you would implement an experiment in which groups can be identified but subjects are anonymous.

2.6. Let us assume that a friend has given you a tiny model of a Mini Cooper. As you examine the amazing detail in this model, you start to think about external, internal, and ecological validity. Explain how the car you have in your hand is viewed in terms of each of the three types of validity.

    a. Is the model like other cars?
    b. Is the model like other Mini Coopers?
    c. Is the model like a real Mini Cooper?

2.7. Design an experiment in which you examine the effect of television cartoons on a child's behavior. If the child watches no television, then what type of behavior would you predict? If the child watches cartoons that depict fantasy situations, then what type of behavior would you predict? Now assume that you have placed a child who has never watched television in one isolated room and a child who has watched a fantasy cartoon in another isolated room. What experiment could you conduct on the children to isolate the effects of cartoon watching?

2.8. Assume that you want to test whether voters tend to vote for a candidate with well-known positions when the position of another candidate is unknown. Consider our judicial candidate race between C and P. Assume that a voter knows that C's rating is 75 but does not know P's rating. Also assume that the voter does not know his own preference on the issue. The question is, assuming no abstention, for which candidate will the voter vote? Design an experiment to test this question by using a script and then design the same experiment in a generic environment. Speculate how the experiments might produce different results.

ORDINAL UTILITY THEORY

3.1. This example is drawn from Roth and Sotomayor (1990). Suppose that a person has preferences such that A is preferred to B, B is preferred to C, and C is preferred to A. That person is in possession of A and you are in possession of C. If you offered this person C for one penny, would he accept? Why? Assume that the person accepts and is now in possession of C, so you offer to exchange him C for B if he will give you a penny. Would this person accept? Why? Now you offer to exchange him B for C for one penny. Would this person accept? Why? Assume that the game ends and explain the outcome for both persons. If you could do this exchange again, would you?

3.2. This example is also drawn from Roth and Sotomayor (1990). Suppose there are four students who must decide on which two of them will be roommates in a dorm. The four students have the following preferences:

Student 1: Student 2 > Student 3 > Student 4
Student 2: Student 4 > Student 1 > Student 3
Student 3: Student 1 > Student 2 > Student 4
Student 4: Student 1 > Student 2 = Student 3

What is the optimal matching scheme for this group of students?
Now consider the following preferences:

Student 1: Student 2 > Student 3 > Student 4
Student 2: Student 3 > Student 1 > Student 4
Student 3: Student 1 > Student 2 > Student 4
Student 4: Student 1 > Student 2 > Student 3

In this example, no one likes Student 4 because she has a reputation of being a slob. Is there an optimal matching scheme? If not, what must happen in order to ensure that the roommates are matched into twos?

3.3. What is the best musical album of all time? To a certain extent, the answer to this question depends on an individual's personal background. For example, an individual's musical taste depends on many factors, such as when a person was born, the type(s) of music to which the person listened growing up, the regions in which the person grew up, his or her socioeconomic background, his or her education level, his or her race or ethnicity, and so on. Hence, when this question is asked, we will get a multitude of responses in which no answer is incorrect but is based on the varied factors outlined here. But what if we really want to know what is considered to be the best album of all time in the aggregate? There are two ways to discover this. Either the best albums of all time can be considered to be the albums that sold the most copies and grossed the most money, or we can ask experts in the music industry to rank the best albums of all time. According to the first criterion, the list of best albums would be:

1. *Thriller*, Michael Jackson
2. *Back in Black*, AC/DC
3. *Bat out of Hell*, Meat Loaf
4. *The Dark Side of the Moon*, Pink Floyd
5. *Their Greatest Hits (1971–1975)*, The Eagles, and *Dirty Dancing*, various artists
6. *The Bodyguard*, Whitney Houston/various artists
7. *The Phantom of the Opera*, Andrew Lloyd Webber
8. *Millennium*, Backstreet Boys
9. *Saturday Night Fever*, The Bee Gees/various artists
10. *Rumours*, Fleetwood Mac

The list generated by experts in the music industry was[1]:

1. *Sgt. Pepper's Lonely Hearts Club Band*, The Beatles
2. *Pet Sounds*, The Beach Boys
3. *Revolver*, The Beatles
4. *Highway 61 Revisited*, Bob Dylan
5. *Rubber Soul*, The Beatles
6. *What's Going On*, Marvin Gaye
7. *Exile on Main Street*, The Rolling Stones
8. *London Calling*, The Clash
9. *Blonde on Blonde*, Bob Dylan
10. *The White Album*, The Beatles

Notice that there is no overlap between the two lists. So what is going on here? Provide an explanation as to why these lists are so different.

3.4. This problem is drawn from Aranson (1981). Consider the following preferences of two people for three political parties:

	Person A	Person B
Utility for communism	1.00	0.00
Utility for fascism	0.83	0.47
Utility for anarchism	0.00	1.00

In this case, a person's most preferred party is fixed at 1.00 and his least preferred party is fixed at 0.00, with the intermediate value u less than or equal to 1.00 and greater than or equal to 0.00. Can we conclude that Person A has a higher utility for fascism than does Person B?

3.5. Consider the following preferences, in which the first in the list is the most preferred, the second in the list is the second most preferred, and the third in the list is the least preferred:

(zxy)   (xyz)   (zyx)   (yxz)   (yzx)   (xzy)

Graphically represent these alternatives using the following dimension. Which of the following alternatives is not single-peaked?

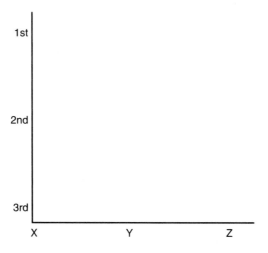

3.6. Consider the following city-block indifference contours, in which x is a person's ideal point. What type of behavior would be consistent with this type of utility function?

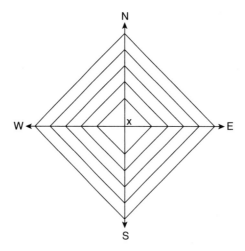

3.7. This problem is drawn from Hinich and Munger (1997). Consider the following two-dimensional utility functions. Salience is the importance that a person attaches to a particular issue. In terms of salience, what is the difference between the two utility functions?

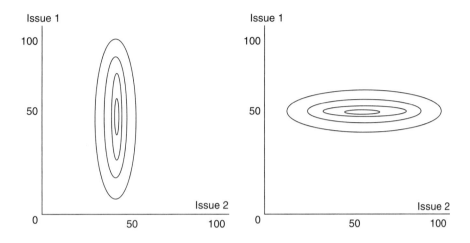

3.8. Consider the following game.

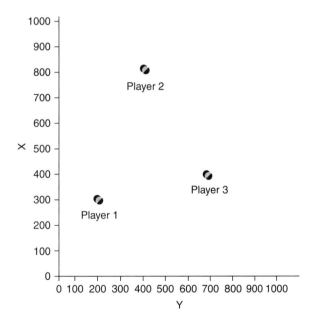

Consider that these three players are deciding between two alternatives, and that each player's preference is represented by a single dot. Each player prefers alternatives that are closer to their dot to alternatives that are farther away. Is there a position within the two-dimensional space that guarantees one alternative a winning position? If so, where is it located?

3.9. Consider a utility function that is based on rationality (maximizing utility) and think of some emotion that you might add to this utility function. Explain how two different emotions such as anger or happiness could change the behavior of a rational player.

3.10. Consider the exercise in question 1.10 and repeat it with the following grouping:

1. Consider two variables: region (South vs. North) and age (young vs. old). Now let us assign numerical values to the variables: South = 1, North = 0, young = 1, old = 0. This gives us four coordinates for each point in a one-unit square (old, south = 0, 1), (old, north = 0, 0), (young, south = 1, 1), (young, north = 1, 0). Within this square, identify areas where issues differentiate the two judicial candidates. Make sure that if one candidate is at 0.25 on one dimension, she should have a similar position on the second dimension.
2. Now specify each election and specify who won.
3. How do these results differ from the results of Question 1.10?

EXPECTED UTILITY THEORY

4.1. Assume that there is a 40 percent chance that you will win 600 dollars and a 60 percent chance that you will win nothing. What is the expected utility for this lottery? If this were a real lottery, how much would you be willing to spend to participate (answers will vary)? Now assume that there is a 90 percent chance that you will win 600 dollars and a 10 percent chance that you will win nothing. What is the expected utility for this lottery? If this were a real lottery, how much would you be willing to spend to participate (again, answers will vary)?

4.2. This problem is drawn from Plott (1976). The Borda voting rule (see Appendix 3) is a voting rule in which alternatives are assigned a numerical weight during the voting process and the alternative with the highest score wins. For example, consider the following simple voting game, in which the most preferred alternative is assigned a value of 2, the second most preferred is assigned a value of 1, and the least preferred is assigned a value of 0:

		2	1	0
	A	x	y	z
Voters	B	x	z	y
	C	z	y	x

In this election, x = 4, y = 2, and z = 3, so x wins.

   Now consider the following game.

		3	2	1	0
	A	x	y	z	w
Voters	B	x	y	z	w
	C	z	w	x	y

Who wins this election? Provide the ranked points.

Now assume that prior to the election, Candidate y dies and is eliminated from the preference profiles. Recalculate the winner using the weights of 2, 1, and 0. Who wins? What condition of rationality does this example violate?

4.3. This problem is drawn from Abrams (1980). Consider the following three individuals with the following cardinal utilities. The winner of the election is the alternative that has the greatest cardinal utility.

		Alternatives	
		x	y
	A	10	5
Individuals	B	7	10
	C	3	4
Sum of utilities		–	–

Now consider an alternate game. Notice that the ordinal rankings are the same and only the cardinal values have changed.

		Alternatives	
		x	y
	A	10	7
Individuals	B	6	10
	C	3	4
Sum of utilities		–	–

Who is the winner in this election? What does this have to say about the worth of cardinal utility? Does this type of election, in which the vote of a voter with a more intense preference counts more than the vote of a less intense voter, make sense?

4.4. This problem is drawn from Abrams (1980). Consider another election in which three voters have cardinal utilities over three alternatives.

Individual A: $x(10) > y(9) > z(8)$
Individual B: $y(10) > z(6) > x(2)$
Individual C: $z(7) > x(6) > y(5)$

Who is the winner of this election if we just add the cardinal utilities for each alternative? Which voters exert more influence over the outcome?

4.5. If you were working for an insurance company and your job was to underwrite policies in which you had to set the price for certain disasters according to a resident's state, how much would you charge residents of New York, Florida, and California for the following types of insurance: flood insurance, hurricane insurance, fire insurance, auto insurance, and earthquake insurance? Rank the five types of insurance in order from the most to least expensive for each state and provide a brief explanation for your ranking.

4.6. At pizza places with kid's entertainment such as Chuck E. Cheese, patrons must exchange real money for tokens to play the various games. In most cases, a token equals a quarter. Because the business has to purchase the fake currency, it seems like it would be more profitable to just use real money. So why does a business make its patrons exchange real money for fake money?

4.7. In simplest terms, what does the rationality assumption mean?

4.8. Las Vegas is the mecca of gambling in the United States. List four ways in which a gambling casino reduces "perceived" risk so that patrons will gamble more money.

4.9. In this chapter, we examined a binary lottery experiment in which the results showed that subjects had a preference for not only how much money a particular subject earned, but also how much money another subject earned. Provide an explanation of why a subject would have preferences of this sort. Why does it matter how much others earn independent of what you earn?

4.10. Think of a situation (this could be a lottery) in which the odds are in your favor but the opponent thinks the odds are in his or her favor. Now think of a situation in which the odds favor your opponent but you think the odds favor you.

SOLVING FOR A NASH EQUILIBRIUM IN NORMAL FORM GAMES

5.1. Consider the following game.

		Player 2	
		Left	Right
Player 1	Up	1, 1	0, –1
	Down	–1, 0	0, 0

What is the equilibrium or equilibria of the game? What is the Pareto optimal outcome?

5.2. Consider the following game.

		Player 2	
		Left	Right
Player 1	Up	1000, 1000	0, 2000
	Down	2000, 0	500, 500

What is the equilibrium or equilibria of the game?

5.3. Consider the following game and solve it by removing dominated strategies. Is this game dominant solvable?

	Player 2				
Player 1	A	B	C	D	E
1	12, 10	11, 11	12, 12	9, 9	8, 10
2	11, 9	10, 10	11, 9	8, 10	7, 9
3	9, 8	10, 7	11, 7	9, 8	8, 7

5.4. Solve the following four games. Which games are dominant solvable?

		Player 2		
		Left	Middle	Right
Player 1	Up	3, –3	23, –23	16, –16
	Middle	1, –1	22, –22	10, –10
	Down	2, –2	21, –21	15, –15

		Player 2		
		Left	Middle	Right
Player 1	Up	11, –11	9, –9	30, –30
	Middle	12, –12	13, –13	27, –27
	Down	10, –10	8, –8	25, –25

		Player 2		
		Left	Middle	Right
Player 1	Up	45, 55	50, 50	40, 60
	Middle	60, 40	55, 45	50, 50
	Down	45, 55	55, 45	40, 60

		Player 2		
		Left	Middle	Right
Player 1	Up	45, 55	10, 90	40, 60
	Middle	60, 40	55, 45	50, 50
	Down	45, 55	10, 90	40, 60

5.5. For the following examples, create two games with strategies and payoffs specified.

a. Consider a case in which you are being robbed by a robber who has a knife. You have the option of giving the robber your wallet or not giving him your wallet. The robber has the option of stabbing you or not stabbing you.

b. Consider two football teams. Team A is on offense and has a better passing game than a running game. Team B is on defense and has a better rushing defense than a passing defense. Assume that the offense can run, throw a short pass, or throw a long pass. The defense can play a run defense, play a pass defense, or it can blitz. If a long pass is selected, the blitz will work more often.

5.6. Reconsider the median voter theorem that we discussed with two candidates and let us assume that there are three candidates competing for some political office. What is the optimal location for each candidate?

Candidate 1	Candidate 2	Candidate 3
Voter 1	Voter 2	Voter 3

5.7. Find the Nash equilibrium for the following three-player game.

		Player 3			
		Right		Left	
		Player 2		Player 2	
		Right	Left	Right	Left
Player 1	Up	12, 9, 6	10, 8, 9	13, 10, 9	10, 9, 10
	Down	10, 11, 12	10, 7, 10	12, 6, 11	10, 8, 13

5.8. Let us assume that Researcher A has constructed a dataset and is uncertain if there are errors in it. Researcher B requests a copy of this dataset. Common professional courtesy dictates that Researcher A should oblige, but in this case, she has incentive to not share the dataset out of fear that the dataset might have errors. Let us consider two situations or states. In State 1, Researcher A has a flawed dataset, and in State 2, she does not have a flawed dataset. Researcher A knows the state but Researcher B does not. Model this situation as two different normal form games: in the first game, Researcher A knows her dataset is flawed, and in the second game, Researcher A thinks her dataset is not flawed. What makes the two games different?

5.9. Consider Company A with a head boss (X) and a second in command (Y). Boss Y sexually harasses one of his employees. The employee wants to report the harassment to Boss X, who can take some action z to sanction Boss Y. However, whether she reports the harassment depends on her perception of the relationship between X and Y and the outcome z. If X is fair, then X and Y are independent and Y does not relate to z. But if X is biased toward Y, then X and Y are related and Y relates to z. One way to think about modeling this situation is to create two games, one in which X is fair and one in which X is biased. In a simple game, the employee's strategies would be to report the charges or to not report the charges, and for X to sanction Y or to not sanction Y. Specify z for each set of actions and model this game. Next, consider this situation one more time and notice the other games that are going on within the basic game, such as a game between X and Y that might involve lying and a game between the employee and Y involving shameful behavior that should be sanctioned.

Model these other two situations and connect the three games by providing some narrative to tell a complete story via the games.

5.10. Let us assume that there is a state of nature that is known by a leader and one of his followers. The leader makes a speech to an audience of followers that describes the wrong state of nature, so the follower who knows the true state accuses the leader of lying. However, the leader then accuses the informed player of lying in front of the audience. Based on these events, who will the other followers believe? Assume that some of the other followers suspect that the leader is a liar. How will this change the outcome? Finally, consider that after the meeting the accuser must decide whether to look the leader in his eyes, and the leader must decide to look the accuser in the eyes. The big assumption here is that the eyes tell the truth. Model this decision in a 2 × 2 game.

5.11. In Question 1.10 you created a simple model of how institutional factors impact the distribution of state supreme court justices along a single dimension. The three institutional factors we considered were voter elections, a partisan legislature, and a bipartisan commission (with you being the dictator in one case). The different institutional factors created different outcomes, so we want to test whether we can create these different institutional factors in an experiment. Consider the McKelvey and Ordeshook (1985) experiment in this chapter and design an experiment to provide such evidence. Keep the following constraints in mind:

   a. Define a single dimension that ranges from one extreme viewpoint to the opposite extreme viewpoint.
   b. Define two judges' positions over the dimension.
   c. Define three voters' preferences over the dimension.
   d. Define a bipartisan position over the dimension.
   e. Define a nonpartisan position over the dimension.
   f. Institute a voting procedure and make information assumptions concerning the players.

CLASSIC NORMAL FORM GAMES AND EXPERIMENTS

6.1. This problem is drawn from Gneezy, Haruvy, and Yafe (2004). Model the following game, referred to as the diner's game.

You and a friend go out to dinner and agree to split the check beforehand. You look at the menu and see the regular meal you would ordinarily order if you were by yourself. However, you notice a more expensive meal that you would not usually order because it is so expensive, but that you know you would like. However, since you will be splitting the check, you reason that this meal will be less costly, since your friend will incur some of the expense. Your friend is also making this same reasoning and debating whether to order a regular meal or a more expensive meal. Model this diner's dilemma, in which both you and your friend must decide whether to order the regular meal or the more expensive meal.

6.2. Create a coordination game in which you and a friend are trying to decide a restaurant at which to eat. You most prefer Thai food and your friend most prefers Mexican food. You both prefer to eat together rather than alone. How would you convince your friend to go to the restaurant you prefer?

6.3. Create a game in which two players have a dominant strategy and the selection of dominant strategies leads to a Pareto optimal equilibrium (deadlock game).

6.4. This problem is drawn from Hirshleifer (1989). Consider two opposing armies during the Hellenic era, of which one must decide to attack by land or by sea, and the other must decide whether to defend by land or by sea. Only consider the payoffs 1 and 2 and create a 2 × 2 normal form game.

6.5. Create a 4 × 4 game in which Player 1 has four strategies (a, b, c, d), Player 2 has four strategies (w, x, y, z), and there is one unique Nash equilibrium to the game. Now create a game with the four strategies and three Nash equilibria.

6.6. Consider the following prisoner's dilemma game.

		Player 2	
		Cooperate	Defect
Player 1	Cooperate	4, 4	2, 5
	Defect	5, 2	3, 3

Let us play a 20-round repeated version of this game. In Game 1, Player 1 alternates between a defect-only strategy and a cooperate-only strategy every five rounds, starting with the defection strategy. Player 2 uses a tit-for-tat strategy. Who wins?

In Game 2, Player 1 alternates between a defect-only strategy and a cooperate-only strategy every five rounds, starting with the defection strategy. Player 2 plays a tit for two-tats strategy, in which he defects only if the other player has defected in both of the previous two rounds. Who wins? Which rule appears to be better in terms of establishing cooperation: tit-for-tat, or tit-for-two-tats?

6.7. This problem is drawn from Arthur (1994). The El Farol bar in Sante Fe offers Irish music on Thursday nights, which is a popular event. The problem is that the bar is rather small. Imagine that there are 100 people considering going to the bar. If less than 60 percent go to the bar, everyone has a great time. But if more than 60 percent go to the bar, it becomes too crowded and a better time would be had staying home. There is no way to determine the number of people who will be there in advance, so a person goes if he expects a turnout of less than 60 percent and stays home if he expects a turnout of more than 60 percent. Model this dilemma as a 2 × 2 game in which the two players are yourself and 60 other players.

6.8. This problem is drawn from Luce and Raiffa (1957). Consider a game in which you and a friend have two strategies: a safe strategy and a double-cross strategy. If both of you play a safe strategy, you both get one dollar. If you both play a double-cross strategy, you both lose five dollars. If one of you plays the safe strategy and the other plays a double-cross strategy, then the one who plays safe gets one dollar and the one who plays double-cross gets one thousand dollars. Model this game. What is its solution?

6.9. What is the Nash equilibrium to a finite prisoner's dilemma game? What is the optimal strategy for both players? Why?

6.10. Tragedy of the commons. This problem is drawn from Hardin (1982). Consider a dilemma in which an agrarian community has a common (or a pasture place) of limited size on which cows can graze. Let us assume that 10 herders who have 10 cows each share this common land, which is suitable for 100 cows to graze comfortably. Here the term comfortable means that milk will be produced at a desired level and the land will be grazed in a manner that will not destroy it. For illustrative purposes, let us assume that our commons is at equilibrium with 100 cows, so that milk is produced at a level of 100 and the quality of the grazing land is maintained at 100. That is, the land is at an equilibrium of (100, 100). Let us also assume that each additional cow increases milk production by 10 but decreases the quality of the commons by 10. This community has no enforcement mechanism to prevent herders from buying new cows instead of just replacing the ones that die. Now let us assume that one of the herders gets a new cow and decides to put it out on the commons. This disrupts our equilibrium, so milk production increases by 10 but the quality of the grazing land decreases by 10, shifting the equilibrium to (110, 90). Now, five herders see the increased milk production and each buy a cow, shifting the equilibrium to (160, 40). Finally, the last four farmers see all the milk that is being produced and each buy a cow, making the equilibrium (200, 0). What does this mean? The grazing land is destroyed, so no more milk can be produced. This situation is called the tragedy of the commons.

    a. What other types of situations can be represented by this dilemma?

    b. What motivates this dilemma?

    c. What type of enforcement mechanisms would maintain our first equilibrium of (100, 100)?

    d. Think about external or internal enforcement mechanisms (e.g., punishment costs, sanctions) and how these might help to resolve this dilemma.

SOLVING FOR MIXED STRATEGY EQUILIBRIUM

7.1. Solve the following:

    a. $6 - 7(x - 3) = -1$

    b. $7(x - 3) = 7x + 2$

    c. $4(x + 1) < 2x + 3$

    d. $(2x - 3/5) \geq x/2 - 1$

7.2. Solve for the MSE for the following games.

		Player 2	
		Left	Right
Player 1	Up	9, −9	10, −10
	Down	10, −10	5, −5

		Player 2	
		Left	Right
Player 1	Up	4, 4	1, 1
	Down	1, 1	2, 2

		Player 2	
		Left	Right
Player 1	Up	–3, 3	1, –1
	Down	0, 0	–5, 5

7.3. This problem is drawn from Haywood (1954). Consider the following example of the Battle of Bismarck Sea during World War II. In this example, the Japanese wanted to reinforce their troops in New Guinea. To get there, they had to go around New Britain on one of two routes: the north route or the south route. The north route generally had bad conditions with rain and winds creating poor visibility, while the south was generally clear. In both cases, the trip would take three days. General Kenney, the commander of the Allied air forces in the southwest Pacific, was assigned to stop the Japanese convoy commanded by General Imamura. He also had two options: to concentrate his forces (aircraft) in the southern part of New Britain, or to concentrate them in the northern part. General Kenney wanted to maximize the number of days he could bomb the fleet, while the Japanese fleet wanted to minimize the number of days it would be bombed. This situation is a zero-sum game, since the best outcome for Kenney is the worst outcome for the Japanese fleet. The problem was to decide which way the Japanese fleet would go. If the fleet went south, it would be easier to find because of the clear conditions (and the greater amount of time they were at sea), and so Kenney would be able to bomb them more. If the fleet went north, Kenney would have less time to find the fleet because of bad conditions. If the fleet went south and Kenney searched north, then it would take a day to discover that the fleet was not there, and Kenney would then go south and immediately find the fleet because of clear weather. If the fleet went north and Kenney searched south, then it would take a day to learn that the fleet actually went north, and it would then take another day to find the fleet because of bad weather. Construct the zero-sum equivalent of this game using the payoffs (3, -3), (2, -2), and (1, -1). Now solve for the equilibrium. If there is no pure strategy equilibrium, solve for the MSE.

7.4. Consider the divide-the-cake game, in which two or more people must divide some limited resource. Let us assume that there are two players who must divide an actual cake.

    a. Describe the fairest procedure that can be used so that the two pieces are equal.
    b. Assume there are three players who must divide the cake. What is the fairest procedure now?
    c. Assume that you are a dictator and can decide how much to give to each player, including yourself. Assume you like each person equally. How would you divide the cake?
    d. Now assume that you are still a dictator but you like one person more than another. How would you divide up that cake? Why?

7.5. In baseball, there is a pitcher and a hitter. The pitcher is trying to throw a pitch to outguess the hitter and the hitter is trying to anticipate the pitch that is being thrown by the pitcher. Let us assume that it is common knowledge that the pitcher has only two good pitches: a fastball and a slider. The batter knows this, so he must try to anticipate whether a fastball or a slider is going to be thrown. If the batter guesses wrong, then she has a strike and receives a payoff of –1, and the pitcher gets a payoff of 1. If the batter guesses correctly, she gets a hit and receives a payoff of 1, and the pitcher gets a payoff of –1. Complete this game and model it as a zero-sum game. Provide a solution to this game.

7.6. This problem is drawn from Moulin (1986). Consider a game in which a father is planning on giving a gift to his daughter. He tells the daughter that he has a gold coin in either his left or his right hand. If she guesses correctly and the coin is in his right hand, he will give her one gold coin, and if she guesses correctly and it is in his left hand, he will give her two gold coins. If she guesses incorrectly, she gets nothing. Construct this 2 × 2 game and if it does not have an equilibrium in pure strategies, solve for the MSE.

7.7. Consider the decision between turning in a really bad paper to a professor and damaging your reputation and not turning in a paper and damaging your reputation in a different way. What are the differences in reputational damage? Model this as a 2 × 2 game in which the professor can impose light reputational damage or heavy reputational damage and you have a choice of turning in a bad paper or no paper.

7.8. Create a game in which a lawyer is considering whether to contribute to a judge's reelection campaign and is playing against a future chance event instead of another player. First consider a binary lottery situation in which the lawyer has a probability 0 of never appearing before the judge and a probability 1 of appearing before the judge. Now assume that there is a 50 percent chance he will appear before the judge. Justify all payoff rankings. What are the equilibria to these games? What assumption is driving these results?

EXTENSIVE FORM GAMES AND BACKWARD INDUCTION

8.1. Solve the following three-person games using backward induction.

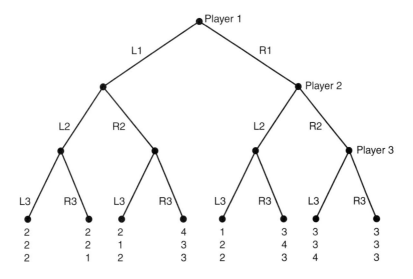

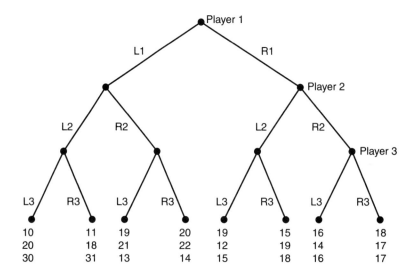

8.2. Solve the following four-person game using backward induction. The payoffs are single digits. Be sure to trace the equilibrium path.

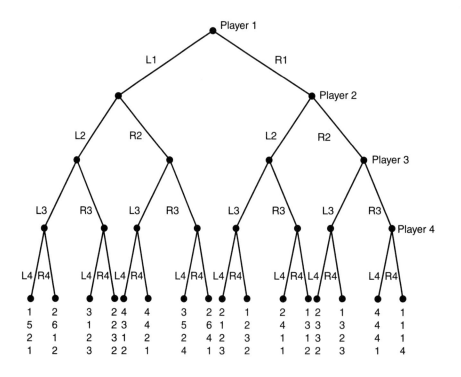

8.3. There are two distinct stimulus packages A and B being debated in Washington. Congress likes Proposal A and the president likes Proposal B. The proposals are not mutually exclusive—that is, either, both, or neither may become law. Thus, there are seven possible outcomes, and the rankings of the two sides are as follows, with a larger number representing a more favored outcome.

Outcome	Congress	President
A becomes law	4	1
B becomes law	1	4
Both A and B become law	3	3
Neither becomes law (status quo)	2	2

The moves in the game are as follows. First, Congress decides whether to pass a bill and whether it will contain A, B, both, or neither. Then the president decides whether to sign or veto the bill. Draw a tree for this game and find the solution using backward induction.

8.4. Let us reconsider our vampire teenage love story. Imagine that our equilibrium prediction is realized and the two marry. Now the couple has to decide whether to have a baby or not, with the probability that the baby will be a vampire being greater than zero. Who would benefit the most from having a vampire baby? Recall that the teenage girl is not a vampire. Model this as a simultaneous move game.

 a. Now assume that the couple decides to have a baby but there are complications during birth and the teenage girl is on the brink of death. The vampire has to decide between letting her die and turning her into a vampire, and the girl has to decide between dying as a human or living as a vampire. Assume that the vampire has the first move. Model this game.

 b. You should now have three games (including the game in Chapter 8). These games have assumed that the vampire is in love with the girl. Now assume that the vampire does not love the girl but needs her to produce a vampire son. Reconstruct the games to fit this story.

8.5. The following is an interpretation of the rivalry between the United States and the Soviet Union for geopolitical influence in the 1970s and 1980s. Each side has the choice of two strategies: aggression and restraint. The Soviet Union wants to achieve world domination, so aggression is its dominant strategy. The United States wants to prevent the Soviet Union from achieving world domination, so it will match Soviet aggression with aggression and restraint with restraint. Specifically, the payoff table is:

		Soviet Union	
		Restraint	Aggression
United States	Restraint	4, 3	1, 4
	Aggression	3, 1	2, 2

a. Consider the game when the two countries move simultaneously. Find the Nash equilibrium.

b. Consider two sequential games: United States moves first and the Soviet Union moves first. Solve using backward induction.

8.6. This problem is drawn from Moulin (1986). Consider the following game of removing sticks. The vertical lines in the figure are sticks. In this game, Player 1 moves first and can remove one or two adjacent sticks, Player 2 can then remove one or two adjacent sticks, and so on. The winner is the player who grabs the last stick. Who wins? Be sure to write out your strategy.

8.7. Draw the extensive form of the prisoner's dilemma game. Solve this game using backward induction. How does this game differ from its normal form representation?

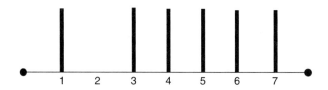

8.8. Reconsider the 21 coin game discussed in this chapter. What strategy could Player 2 use to guarantee herself a win, assuming that Player 1 does not know the optimal strategy?

8.9. This problem is drawn from Davis (1982). This is a game of simplified poker with two players, A and B. Each player places five dollars in the pot. Each player also has a two-sided coin with a 1 on one side and a 2 on the other, which they throw prior to game play. Both coins are placed in a position where neither player can see what the other player has thrown. Player A has the option to fold or bet. If Player A folds, then the coins are examined and the player who has the highest number gets the 10 dollars in the pot. If the numbers are the same, then both players get their money back. However, if Player A bets, she must place three more dollars in the pot. Player B can then either fold or see. If Player B folds, then Player A wins the money in the pot, regardless of the coin toss. If Player B sees, then she must place three dollars in the pot and whoever has the highest number on their coin wins the pot. Again, if the numbers are the same, then each player gets his or her money back. Draw the game tree for this poker game.

SUBGAME PERFECT EQUILIBRIUM

9.1. This problem is drawn from Watson (2002). How many subgames are there in this game?

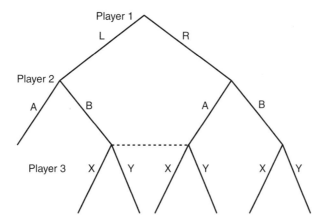

9.2. What is the subgame perfect equilibrium for the following game?

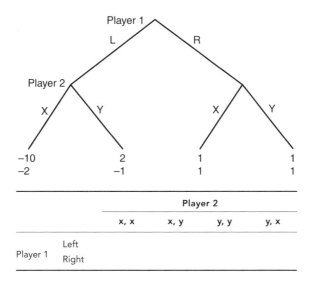

		Player 2		
		x, x	x, y	y, y	y, x
Player 1	Left				
	Right				

9.3. Consider the following normal form game.

		Player 2	
		Left	Right
Player 1	Up	−1, 3	0, 4
	Down	0, 2	−1, 1

a. Identify the equilibria or equilibrium for this game.

b. Draw the extensive form for the game, with Player 1 moving first.

c. Solve for subgame perfect equilibrium (write out the 2 × 4 game) and identify all the equilibrium or equilibria for the 2 × 4 game. Identify any subgame perfect equilibrium or equilibria that exist.

9.4. Consider the following extensive form game.

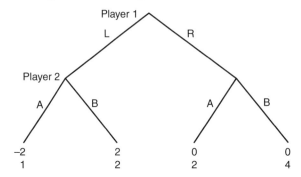

a. Write out the 2 × 2 normal form game and identify the equilibria or equilibrium.

b. Solve for subgame perfect equilibrium (write out the 4 × 4 game). Identify the subgame perfect equilibrium.

9.5. Consider the following four cities:

1. Ciudad Juárez, Mexico (one of the most violent cities in the world)
2. Flint, Michigan (one of the most violent cities in America)
3. Amsterdam, the Netherlands (one of the safest cities in the world)
4. Newton, Massachusetts (one of the safest cities in America)
    a. Use one of the bargaining models discussed in this chapter and explain how the environment of each city could impact bargaining outcomes.
    b. Consider a relief scale that measures the relief someone feels after he or she has made it back home after doing routine chores in the city. List four cities in your state or region and place them on a similar relief scale. How would the bargaining model you discussed previously be impacted by a location in these four different cities?

9.6. Let us consider that two people have fallen in love but each lives in a faraway country. However, each individual most prefers to stay in his or her own country. Construct this scenario as a game in which the two people must decide whether to stay in their own country or to leave their country. Who makes the first decision? Identify and explain preference orderings for the various contingencies. Identify a subgame equilibrium for the game. Give a narrative about what happens in your game and identify the major assumption driving the outcome. Explain how this game changes if you assume:

1. Cities in a state instead of different countries.
2. The two people intend to marry.
3. One person has a greater income than the other.
4. One person loves his or her country more than the other.

9.7. This problem is drawn from Charron (2000). In the fifth century B.C., the Athenians controlled a large portion of the Aegean and assumed the responsibility of defending the Greek world against Persian invasion. Sparta had neither the will nor the ability to assume such a role. After the defeat of the Persians, Athens had to decide whether to continue maintaining the empire (e), with all of its attendant costs, or to allow it to dissolve of its own accord (d). The mere existence of the empire provoked hostility in the Spartans, who feared Athenian imperial power. However, a decision to dissolve the empire would pose a security risk to Athens, as any of the satellites states could then shift their allegiance to Sparta (s)—not an unlikely outcome, given their resentment toward Athens for the city's high taxation rates and violent suppression of local political innovations. Given the options, concerns, and resulting preferences of each party, model this three-player version of the centipede game and solve for the subgame perfect equilibrium.

Outcome	Sparta	Athens	Small cities
War	2	2	2
Athens retains empire	1	4	1
Small cities join Sparta	4	1	4
Small cities stay independent	3	3	3

9.8. Two eight-year-old girls K and M are deciding whether to annoy an adult supervisor or to play among themselves. K most prefers to annoy an adult and M most prefers to play with K. Specify a normal form game for this situation and justify all payoff rankings. Model as an extensive form game with K moving first. Solve for the subgame perfect equilibrium. Redo the analysis assuming that K and M most prefer to annoy an adult.

9.9. Consider a congressional conference committee that is responsible for making the Senate and House versions of a bill exactly alike in order for it to be passed and signed into law by the president. Assume that Congress is considering judicial term limit reforms. The question is whether federal judges should have life-time tenure or should be subject to some limit such as age 79 or 20 years of service. The issue is accountability. The Senate assumes a more traditional approach, preferring that judges be accountable to their personal doctrines and follow their own beliefs. The House prefers that judges be accountable to the public, or that laws change with the times. List all the bargaining options that the Senate and House would have to debate. Is there a compromise position? What assumptions would have to be made so that the Senate would have a bargaining advantage and be able to achieve an outcome closer to their position? Repeat this analysis and list the assumptions whereby the House would have a bargaining advantage.

IMPERFECT AND INCOMPLETE INFORMATION GAMES

10.1. Create the following three-person game. In this game, Player 1 does not know her type and has two choices at each of her two nodes. Player 2 knows Player's 1's type but does not know which move she has made. Player 2 has two choices at each of his four decision nodes. Player 3 knows Player 1's type but does not know how Player 2 has moved.

10.2. Give an interpretation for the following game. What does Player 2 know, what does Player 3 know, and what does Player 4 know?

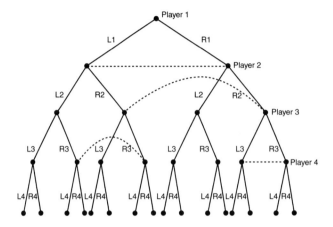

10.3. Solve for the sequential equilibrium for the following game.

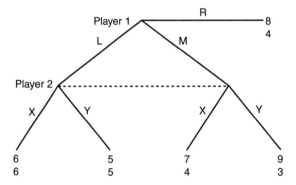

10.4. Solve for the sequential equilibrium for the following game.

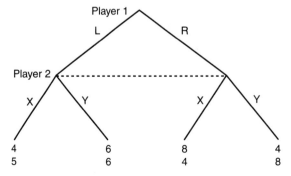

10.5. In this game, Player 1 has the first move and has the option of passing the play to either Player 2 or Player 3 by selecting A. If she passes to Player 2, he has the option of playing C and ending the game or giving a move to Player 3. If Player 3 gets a move, he does not know if Player 1 or Player 2 passed the move to him. What is the sequential rational outcome of this game? Hint: You only need to calculate Player 3's expected utility.

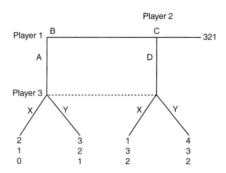

10.6. Think of three non-Disney movies and explain how asymmetric information is incorporated into the story's script.

10.7. Create an extensive form tree in which two players know each other's move in part of the tree, but have a simultaneous choice move in another part of the tree. Now create an extensive form tree in which Player 1 moves twice and forgets her first move and Player 2 moves twice and forgets her first move.

10.8. Solve the following signaling games.

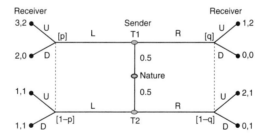

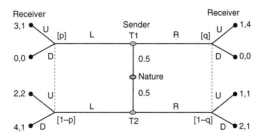

10.9. Solve for the sequential equilibrium for the following game.

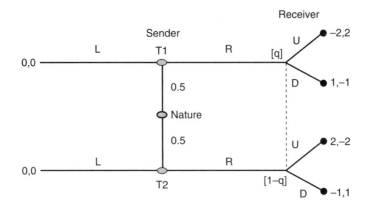

10.10. Construct the following political game. A politician has been elected to some public office and has the option of hiring a first-class staff (experts) or a second-class staff (nonexperts). She is constrained by money, so that if she hires a first-class staff, she will have to do without the perks normally given to a politician (e.g., perks unobserved by typical voters, such as a new oak desk). The politician is the sender and can be of two types: low ability or high ability. A voter is a receiver who either votes for the candidate or does not vote for the candidate. Voters prefer high-ability candidates to low-ability candidates and will vote for a candidate if they perceive her to be a high-ability type. The sender wants to be perceived as a high-ability candidate so that voters will vote for her. Hence, if she has low ability, she wants to hire a first-class staff, and if she has high ability, she wants to hire a second-class staff.

    a. Construct a game in which the candidate's type is not revealed to the voters.

    b. Construct another game in which the candidate's type is revealed to the voters.

10.11. *Truth game.* Consider a game in which a person is applying for a job. The employer has two job openings: one that is demanding and one that is less demanding. The demanding job requires a high-ability type and the less demanding job requires a low-ability type. The applicant prefers the less demanding job if she is low ability and the demanding job if she is high ability. In this game, nature selects an ability level for the applicant (low or high ability). This applicant sees her type and must decide whether to tell the truth and report her ability level or lie and report the opposite type that was revealed to her. Remember that the applicant will be happiest with the job that corresponds with her ability level. Construct this game.

Lying game. Reconsider the game you just constructed and now assume that the applicant wants the demanding job regardless of her type. In this case, she must lie if she is low ability. Construct a version of the lying game.

10.12. In John Maynard Smith and David Harper's book *Animal Signals*, the authors give the following example of mutual display, which is when two or more animals display actions simultaneously in an apparent attempt to signal something. They note (2003, 127):

In some cases, the occasion for such a group display is still obscure. For example, in the marching display of flamingos, a densely packed group of birds marches very fast, abruptly reversing direction at intervals....The plumage is ruffled, increasing its apparent pinkness, and the birds jerk their heads from side to side....This behavior is most developed in the Lesser Flamingo, in which marching groups can contain hundreds or even thousands of birds....Marching is often interpreted as a means of synchronizing breeding, and pairs of birds do sometimes separate from the group and display with each other. There is, however, no temporal association between marching and nesting…in contrast with other group displays which are most frequent in Great Flamingo during the last 2 months up to egg-laying....Similarly, suggested association between marching and other behaviours such as feeding or dispersal are not supported by data. It therefore remains unclear what function marching serves.

Speculate why the flamingos behave in this manner and what signal they are trying to convey.

10.13. This problem is drawn from Cho and Kreps (1987). Consider the beer-quiche game. In this game, two players are deciding whether to duel with swords. Whether or not they duel depends on what signal the receiver sends. The sender can be one of two types: wimpy or surly, with wimpy types hating to duel and surly types loving to duel. In this instance, the receiver will agree to duel if the sender's type is wimpy but will be reluctant to duel if the sender's type is surly. If one player backs out of the duel, he will be humiliated and will receive less utility than the other player. The sender eats breakfast before the duel and can either drink beer or eat quiche, which sends a message to his opponent (obviously only surly types drink beer and weak types eat quiche for breakfast). Model this signaling game assuming that the probability of the sender being surly or weak is 0.5.

10.14. Consider a game in which the mayor of a Red city has been accused of racial discrimination against the Blues in the city. There are two states of nature that can occur, one in which the mayor is a bigot and another in which he is not a bigot. The mayor has the option of passing some policy that will hurt the Blues or passing a policy that will help the Blues. The Blues can either protest the mayor or not protest the mayor. Construct a game in which the Blues do not know if the mayor is a bigot or not. Construct payoffs so that a story can be told that makes sense and provide a narrative that describes the story.

BAYESIAN LEARNING

11.1. Think of events that could occur that could change your current beliefs about:
    a. The creation of life
    b. The political party you dislike the most
    c. Your favorite sporting team and/or singer
    d. Your mood doing this exercise
    e. What you think you will do when you get home

11.2. In 2010, astronomers discovered a planet that was a similar size as Earth and was orbiting around a red dwarf star called Gliese 581, 20 light years away. Astronomers believe that this is a possible habitable planet with enough gravity to retain an atmosphere. Let us imagine that in the distant future, people from Earth will want to inhabit this planet. In order to live on this planet, people from Earth will have to develop a completely different belief system. About what types of things might the new inhabitants have to develop new beliefs?

11.3. Explain what weak consistency of beliefs means.

11.4. Describe two independent states and different actions that two players take within each of the states. Explain why the states you have selected are independent. Also explain the type of behavior that changing states created for you.

11.5. Consider the following strategy choices for Player 1 and Player 2.

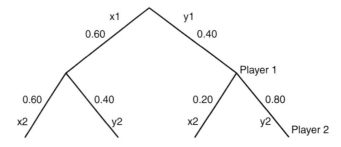

    a. Calculate the conditional probabilities for each of the four cases.
    b. What is the posterior for each of the four cases?

11.6. Consider the following two-player game.

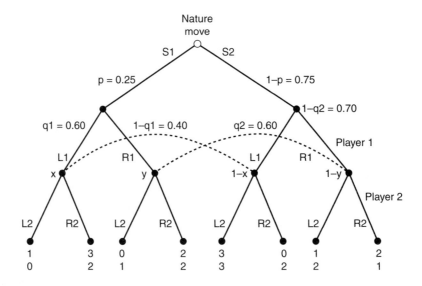

a. Calculate Player 2's beliefs (x and y).
b. Solve for the perfect Bayesian equilibrium.
c. Assume that p = 0.75. What is the perfect Bayesian equilibrium now?
d. Did changing the state of nature change the equilibrium?

11.7. Think of a real-world case in which beliefs off the equilibrium path have impacted your behavior. That is, someone took an action that affected your payoff, but you had no choice in the matter. For example, assume that your boss refused to give you a raise on the grounds that there were no raises to give, but you know for a fact that other people did get a raise. In this case, your boss took an action that affected your payoff. You have beliefs (you believe that your boss is a liar), but since you have no action to take, it is unclear what you should do with those beliefs. This belief may also further diminish any payoff that you receive in this equilibrium. Provide another example in which off-the-path behavior affects behavior in a real-world situation.

11.8. Consider the following game.

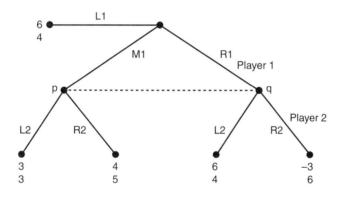

a. Draw the normal form for this game and identify its Nash equilibria.
b. Informally describe why one of the equilibria is a perfect Bayesian equilibrium and the other is not (or maybe should not be).

11.9. Consider a game in which Olga is trying to decide whether to give a gift to Igor or not. As we know, Olga can be of two types: a friend of Igor or an enemy of Igor. As a friend, Olga has a nice gift in her purse, and as an enemy, she has a bad gift in her purse. The gifts are wrapped, so Igor has to decide whether to accept the gift without knowing what is in the gift package. In this game, Olga has the option to either give the gift (G) or not give the gift (DG), and Igor has the option to either accept (A) or reject (R) the gift. The payoffs are structured so that if Olga is a friend, Igor wants to accept the gift, and if Olga is an enemy, Igor wants to reject the gift. Olga prefers for Igor to accept the gift rather than reject it, and if she thinks Igor will reject it, she prefers not to give the gift.

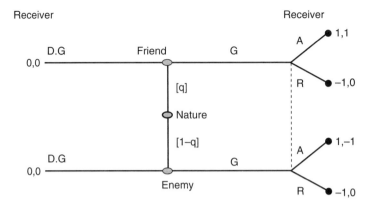

If no gift is given, the payoffs to Olga (the sender) and Igor (the receiver) are just zero. The best case is for Olga to be a friend and for Igor to accept a nice gift, in which case they get a payoff of (1, 1). If Olga is a friend and Igor refuses the gift, then Olga's feelings are hurt and Igor is unaffected, so their payoff is (–1, 0). If Olga is an enemy and Igor accepts her gift then she gets satisfaction from giving a gift but Igor gets a terrible gift, so their payoffs are (1, –1). Finally, if Olga is an enemy and Igor rejects the gift, then Olga is disappointed and Igor is unaffected, so their payoffs are (–1, 0). Use the tools we have learned in the last two chapters to solve this game.

11.10. A popular theory in sociology is social network theory. In this theory, the way that a person behaves is a function of the social network of which he or she is a part. For example, if you are on Facebook or some other social networking site, then your social network would consist of the friends you have befriended on Facebook. It is argued that a flaw of rationality is its failure to consider this social network behavior, because these networks (and the configuration of connected networks) impact behavior. In terms of social networking, to what aspect of rationality does social networking relate—utility functions, payoffs, beliefs, or strategies? What is the primary way in which social networking would impact players in a game environment? If, for example, you selected beliefs, then how does social networking affect beliefs in a model? And how might social networking be incorporated in models? Think about information overload, filtering of information, communication among players, and so on. Finally, is this area of study similar to computational behavioral game theory?

11.11. This problem is drawn from Samilton (2011). In 1994 General Motors and Ford were bargaining with Chinese officials to replace their Chinese-made limousines with Western-style cars. At the time, most Chinese officials wore Western-style hats. GM knew about the hat culture but Ford did not. Ford presented a small Taurus to the officials and GM presented a big Buick. When a Chinese official climbed into the Taurus, his hat fell off because of its small size, but when he got into the big Buick, his hat stayed on. GM got the contract. Model this situation as an asymmetric bargaining game in which Ford knows about the hat culture and GM does not. Now assume this information structure and consider a decision by GM officials to present a small car or a large car.

NOTES

1. http://www.rollingstone.com/news/story/5938174/the_rs_500_greatest_albums_of_all_time

SOLVING LINEAR EQUATIONS

1. Find value for the variable x:

$$10 - 3x = 7$$
$$10 - 3x(-10) = 7(-10)$$
$$-3x = -3$$
$$x = 1$$

2. Find value for the variable x:

$$2(x + 5) - 7 = 3(x - 2)$$
$$2x + 10 - 7 = 3x - 6$$
$$2x + 3 = 3x - 6$$
$$2x + 9 = 3x$$
$$x = 9$$

3. Consider the following equations with fractions: $5/4x + \frac{1}{2} = 2x - \frac{1}{2}$

   To get rid of the fractions, multiply both sides by the lowest common dominator (LCD) of 4.

$$(4)(5/4x + \frac{1}{2}) = (4)(2x - \frac{1}{2})$$
$$5x + 2 = 8x - 2$$
$$5x + 2 - 8x = 8x - 2 - 8x$$
$$-3x + 2 = -2$$
$$-3x + 2 - 2 = -2 - 2$$
$$-3x = -4$$
$$-3x/-3 = -4/-3$$
$$x = 4/3$$

4. Solve the following linear inequalities for the variable x:

$$x - 7 < -3$$
$$x - 7 + 7 < -3 + 7$$
$$x < 4$$

5. An inequality stays the same throughout a problem. Adding or subtracting values to both sides does not change the inequality.

$$x + 10 \geq 5$$
$$x + 10 - 10 \geq 5 - 10$$
$$x \geq -5$$

$$-x/2 > 7$$
$$(-2)/(-x/2) < (-2)(7)$$
$$x < -14$$

6. The inverse of dividing by –2 is multiplying by –2, so the inequality sign reverses.

$$-3x \leq 9$$
$$-3x/-3 \geq 9/-3$$
$$x \geq -3$$

$$-3x - 3 < 6$$
$$-3x - 3 + 3 < 6 + 3$$
$$-3x < 9$$
$$-3x/-3 > 9/-3$$
$$x > -3$$

$$4x - 1 = 4(x + 3)$$
$$4x - 1 = 4x + 12$$
$$4x - 1 - 4x = 4x + 12 - 4x$$
$$-1 = 12$$

Where did x go in this equation? The initial equation was false, and there is no solution.

7. To solve the following equations with decimals, we must first get rid of the decimals, so we multiply both sides by 100, and then get all y terms on one side.

$$0.35y - 2 = 0.15y + 1$$
$$100(0.35y - 2) = 100(15y + 10)$$
$$35y - 20 = 15y + 10$$
$$20y - 20 - 15y = 15y + 10 - 15y$$

8. The inverse of subtracting 20 is adding 20.

$$20y - 20 + 20 = 10 + 20$$
$$20y = 30$$

The inverse of multiplying by 20 is dividing by 20.

$$20y/20 = 30/20$$
$$y = 3/2$$

A SHORT HISTORY OF GAME THEORY AND POLITICAL ECONOMY EXPERIMENTS

1. ORIGINS

Game theory's history is fascinating and its evolution is mixed with the history of physics and other sciences as a result of the axiomatic revolution that is the basis for much of science today.[1] I think of game theory as stemming from two primary roots. The first root is the theory of chance (probability theory), whose history began in 1654, when a letter between the French mathematicians Pierre de Fermat and Blaise Pascal first discussed risk in terms of probability theory and expected value theory.[2] The other root is David Hilbert's 1899 book *Grundlagen der Geometrie*. Leonard (2010, 44) notes:

> Hilbert saw the axiomatic approach as providing clarity and rigour. The ideal he had in mind was to establish, for the field under scrutiny, the set of basic postulates that were both necessary and sufficient to generate the laws of the domain. By ensuring that such axioms were independent, there was no redundancy in them. By ensuring their consistency, they could not be a source of contradictory results.

Probability theory first met axiomatic theory in Heisenberg's 1927 uncertainty principle, which specified that the laws of quantum mechanics (axiomatic laws) were probabilistic and could only be examined by statistical predictions. Game theory is an attempt to model social behavior in a "quantumlike" way. Behavioral game theory attempts to explain some randomness in tests of axiomatic models by psychological factors.

Game theory is thus the product of a combination of the axiomatic approach with probability theory. One implication of the axiomatic approach is the existence of a choice function defined by set theory. This choice function allows the examination of the interrelatedness of decisions and choices of two (or more) players within a matrix—like the two-person zero-sum game of chess.

2. TWO-PERSON ZERO-SUM GAMES

In his 1926 book *Lasker's Manual of Chess*, master chess player and mathematician Emanuel Lasker broached the idea that social and political situations could be modeled as parlor games (Lasker 1947). The idea is that if chess is like real life, then we can reduce the game to eliminate any psychological effects and focus on the choices of strategies in order to find an equilibrium. And if these games are solvable, then scientific games modeling human behavior can be created and studied like games of chess. If a two-person zero-sum game can be solved, we can examine decision-making in many different institutions axiomatically and make predictions about outcomes. Ernest Zermelo (1913), also a noted chess player and mathematician, made the important find that in finite games like chess, given a player's starting position, a player can guarantee himself a win. Colman (1982, 11) explains:

> Zermelo managed to prove that every strictly, two person game of perfect information possess either a guaranteed winning pure strategy for one of the players, or guaranteed drawing pure strategy for both. This result applies, for example, to zero-sum board games like chess in which each player knows at every move what moves have been made previously, but it provides no method for finding the winning (or drawing) strategies.

Émile Borel (1938) found a partial solution to this two-person zero-sum game, proving that a person could guarantee a tie if probabilities were considered. Famously, he also once claimed that this game would never be solved (Leonard 2010).

John von Neumann, a child prodigy and student of Hilbert, eventually solved this problem. As a professor at Princeton University, he published an article entitled "The Theory of Parlor Games" ("Zur Theorie der Gesellschaftsspiele") in 1928, in which he created an axiomatic game "as a formal mathematical entity, complete with discussion of the completeness, freedom from contradiction, and independence of axioms" (Leonard 2010, 234). He used this game structure to prove the minimax theorem, which specified that every two-person zero-sum game has a solution if you assume that players' choices are probabilistic.[3] He couched his solution in cooperative game theory, arguing that players are actually two coalitions of people, with side payments or transferable utility involved. For each coalition, there is a value (expected value) to the game based on the different strategy choices. The coalition tries to maximize this value by selecting the strategy with the highest value, but since the game is zero-sum, the opponent tries to select a strategy that minimizes this value. Hence, a solution to the game occurs when both players (coalitions) select strategies in which they attempt to maximize their value while minimizing their opponent's value. This is referred to as a split-the-difference principle.

3. COOPERATIVE GAME THEORY

von Neumann (1928) thought that the appropriate level for these games was the group level, so he began the study of cooperative game theory, in which players can openly communicate and form coalitions with each other. Coalitions are formed when two or more players coordinate their strategies. Consider three coalitions: Coalition 1, Coalition 2, and

Coalition 3. Superadditivity means that if Coalition 1 has a payoff of 3 and Coalition 2 has a payoff of 7, then their combined payoff is 10, and any distribution of 10 among the two coalitions is allowable (unless exogenously restricted). Transferable utility or side payments allow coalitions to split payoffs any way they want (assuming these payments are money, which is linear). Consider a decision rule in which the outcome is determined when two coalitions join a larger coalition in which they split total utility and a third coalition gets nothing. Assume that utility is some value that each coalition has, so Coalition 1 is worth 7 utiles, Coalition 2 is worth 6 utiles, and Coalition 3 is worth 7 utiles. Hence, Coalition 1 could offer Coalition 2 one utile to combine their two coalitions, creating a coalition worth 13 utiles. Coalition 1 could openly communicate with Coalition 2 and make the public announcement, "I will give you a utile if you join my coalition." Although if acted upon, this offer would decrease Coalition 1 by 1 utile, it would spare the coalition the worse fate of not being a member of a larger coalition and receiving nothing. Of course, Coalition 3 might hear the public message and offer Coalition 2 two utiles, and so on. In this case, offers (or bargains) are the result of trying to prevent a player's worst outcome from occurring.

Using axiomatic theory and probabilistic choice functions, the minimax theorem proved that every two-person zero-sum game has an equilibrium in mixed strategies. Because this is such an important development for game theory, I will illustrate a noncooperative version of this game that is consistent with my usage in this book. Recall that players want to minimize the maximum damage that an opponent can inflict. Using this approach, a player adopts a "security level," which is the minimum payoff that a strategy affords if another player chooses the most damaging response. Consider Table A.1.

For now, ignore the last row and column in the table. Consider Player 1's U strategy, whose elements are (3, 5, 4). For this strategy, his security level (or the lowest payoff) is 3, so let us put this number in the last column, which displays Player 1's security levels. Player 1's D strategy lowest payoff is –5, so let us place that number in the last column. Player 2 has three strategies, for which the lowest payoffs are placed on the bottom row. Now we can select the best of the worst payoffs for each player. Player 1's worst payoff is 3 and Player 2's worst payoff is –3. If we examine the cell that contains these two payoffs, we will notice that this cell is an equilibrium, since no player has a positive incentive to defect from it. A player in a two-person zero-sum game chooses the strategy that maximizes her security level, because she assumes that her opponent will select the strategy that minimizes her payoff. Hence, maximizing security levels leads both players to a pure strategy equilibrium if it

TABLE A.1   CALCULATING SECURITY LEVELS

Player 1	Player 2			
	R	M	L	Security level for 1
U	3, –3	5, –5	4, –4	3
D	–5, 5	9, –9	6, –6	–5
	–3	–9	–6	sec. for 2

exists and a mixed strategy equilibrium otherwise. This strategy produces a number v, which is the best payoff the players can ensure themselves. Camerer (2003, 29) notes:

> Mixed-strategy equilibrium is a curious concept. Introducing mixed strategies makes the space of payoffs convex (i.e., for any two points in the space, all points in between are in the space too), which is necessary to guarantee existence of a Nash equilibrium (in finite games). Guaranteed existence is a beautiful thing and is part of what makes game theory productive: For *any* (finite) game you write down, you can be sure to find an equilibrium. This means that a policy analyst or scientist trying to predict what will happen will *always* have something concrete to say.

Von Neumann's treatment of this game as a cooperative game began the field of cooperative solution theory, which Ordeshook (1986, 387–88) explains:

> Cooperative solution theory...focuses on subsets of outcomes and on the properties that those subsets satisfy. Thus, we predict a specific outcome not because of its properties alone, but because this outcome is a member of some set and that set satisfies certain properties, properties that it might not satisfy if that outcome is excluded. Our predictions no longer take the form "the outcome *o* will prevail because *o*...[defeats, ties, is greater than or dominates]...all other possibilities," but rather "*o* is a possible outcome because it is a member of the set *X*, and *X* satisfies the properties..., which *X* would not satisfy if *o* is excluded from it." Hence, this approach admits the possibility that we can only limit predictions to sets of outcomes. And to the extent that such a set contains more than one element, this approach admits of social processes that we can characterize by fundamental indeterminacies.

In short, cooperative solution theory is concerned with conditions or institutional structures that create these subsets of outcomes.

4. ECONOMIC EXPERIMENTS

After the two-person zero-sum game had been solved, it was time to test the theory, and experimental methods were the primary method of testing these axiomatic models from their inception. The first person to systematically use experiments in the social sciences was the Russian physiologist and Nobel Prize winner Ivan Petrovich Pavlov, who, during the period of 1891 to 1900, conducted experiments that aimed to explain conditional reflex. Famously, Pavlov conditioned dogs to salivate at the sound of a bell, which had previously been associated with the sight and smell of food.

In the 1920s and early 1930s, the utility function, or preferences of players (which were assumed from the axiomatic choice theory), became an increasing focus of interest. In 1929, Harold Hotelling published his article "Stability and Competition," which was concerned with nonconvex preferences that cause consumer demand to not be connected. Experiments on individual choice were first conducted by Louis Leon Thurstone, who was an assistant to Thomas Edison in 1912 and later went on to develop the statistical technique of multiple factor analysis. In 1931, he authored a paper entitled "The Indifference Function," which was concerned with testing individual indifference curves. Indifference curves are the standard way that economists model individual preferences. Thurstone conducted experiments in which subjects made hypothetical choices among commodity bundles such as shoes and hats, hats and coats, or shoes and coats. His results supported the validity of indifference curves. It is interesting that in 1959 he commented:

> The paper on the indifference function was the outgrowth partly of my sincere belief that economics could and should be an experimental science. I have found that my colleagues in this field are divided on this question. Some of them deny emphatically that economics could be an experimental science.

This initial experiment was criticized on the grounds that the choices to which subjects responded were strictly hypothetical, and because no real stimuli were involved, the responses were valueless (Wallis and Friedman 1942).

5. NONCOOPERATIVE GAME THEORY

Game theory's foundations were not fully constructed until the publication of von Neumann and Morgenstern's seminal *Theory of Games and Economic Behavior* in 1944. This book, which has the reputation of being the most influential book that only a few people have actually read, was their attempt to build a social theory based on axiomatic principles. Critically, it developed the assumptions of expected utility theory.

The next major development in game theory was the establishment of the Nash equilibrium for non-zero-sum games, conceived by Princeton professor John Nash.[4] Nash proved that an equilibrium exists in a game without coalitions (which von Neumann had to assume).[5] This equilibrium concept was based on the question that if a player played a game once and observed the outcome of game play, and then was given the opportunity to play the game again, would this player change the way she played the game? A Nash equilibrium is an outcome in which no player has an incentive to change his or her decisions or has any regret over the strategy he or she selected. As in von Neumann's minimax theorem, Nash's equilibrium was solved in terms of mixed strategies. The importance of the Nash equilibrium is that it diminished the importance of coalitions in the analysis and began the development of noncooperative games, which are the major focus of study today.

Many people do not understand the importance of John Nash in terms of laboratory experimental methods. In fact, Nash conducted the first noncooperative game theory experiment on his bargaining game solution (which consists of a set of assumptions related to fair outcomes). Dimand comments (2005, 9):

> Kalisch, Milnor, Nash and Nering played six constant-sum, cooperative games with side-payments: four were four-person games played eight times each, the others a five-person game played three times and a seven-person game played twice. Players were rotated to discourage permanent coalitions. They found what they considered "a reasonable good fit between the observed data and the Shapley value" [equilibrium] but found that results differed between games that were strategically equivalent and that symmetries between games were not well reflected in observed outcomes. They also

played a three-person cooperative game with no side payments, and attempted to ascertain the workability of a negotiation model, defined as "a non-cooperative game based on a strictly formalized negotiation procedure applied to a cooperative game"…the solution to which could be considered as a possible solution to the underlying cooperative game.

6. THE RAND CORPORATION

Against a backdrop of postwar fear over a nuclear attack by the Soviet Union, the RAND corporation was established by the federal government in Santa Monica, California, in the 1950s so that the great thinkers of the time could meet to combat the Soviets. Scientists from all fields gathered to think and theorize about effective combat strategies. Under the leadership of John Williams, the game theory division within RAND became an important component of the organization, since its focus aligned with war strategies in which two players act as combatants and each side is trying to maximize damage given but minimize damage taken (Leonard 2010). Although von Neumann was at RAND at the time, he was already thinking about the design of computers (theory of Automata). It was Lloyd Shapley who applied the two-player zero-sum structure to military applications.

In 1954, RAND held a conference with the leading game theorists and experimentalists and published the resulting papers (Thrall, Combs, and Davis 1954).

Two other RAND scholars, Merrill Flood and Melvin Dresher, devised the prisoner's dilemma game as a result of an errant car deal by Flood, in which two players have an incentive to not cooperate when it is in their best interest to do so (Flood 1958).[6] This simple game greatly influenced the spread of game theory, since it was easy for non-mathematicians to understand and had important applications in the fields of political science, economics, law, biology, sociology, and psychology. A group of RAND researchers played the game 100 times in the game's first experiment. As opposed to the solution derived by the Nash equilibrium that predicted defection, the experiment found quite a bit of mutual cooperation. When John Nash was asked about the outcome, he commented that if scholars wanted to test his equilibrium solution, then each game should be considered independently of the other and not as a repeated game, as in that experiment, in which payoffs are a function of each successive game (Poundstone 1992). His point was that it is necessary to control for repeated game effects when the strategy choice of one period is a function of strategy choice in another period—that is, the experiment needs to control for repeated game effects. What is important about this experiment is that it created a large number of other experiments in all disciplines that attempted to solve the dilemma.

There was some dissatisfaction at RAND that although game theory made solid predictions, no experimental verification was available. Leonard (2010, 329–30) recounts:

Flood insisted that game theory was mathematically rigorous and of great value, but of questionable validity insofar as it had not been shown to stand up to experimental test. Even two-person zero sum theory, the most reliable part, said nothing about how a player would actually play in even simple games. In "Matching Pennies," it predicted the use of randomization but said nothing about how a player might use observations of his opponent's past play, something which even a child could be expected to do.

Experiments on industrial organization began in 1948 when Edward Chamberlin, a professor of economics at Harvard, conducted classroom experiments to test his theory of supply and demand, which was designed to support his earlier theory of monopolistic competition (Chamberlin 1933). The experiments were conducted without monetary incentives, students did not take them very seriously, and Chamberlin did not find any empirical evidence for his predictions (see Dimand and Dimand 2005). However, a first-year graduate student at Harvard, Vernon Smith, did take them seriously, and in 1956, he began running Chamberlin's experiments with real monetary incentives. Unlike Chamberlin, Smith found convergence to the equilibrium prediction, publishing his results in 1962. In 2002, Smith won the Nobel Prize for his contributions to developing experimental economics.

7. GAME THEORY EXPERIMENTS

There is no doubt that game theory experiments have been influenced by B. F. Skinner's 1938 text, *The Behavior of Organisms*, which began the behavioral revolution that is essentially concerned with the experimental study of human behavior. The link between psychological experimental methods and economic experiments is illustrated in Reinhard Selten's auto-biographical statement written for the Nobel Prize committee:

> My first publication was a journal article with the title "Ein Oligopolexperiment" (an oligopoly experiment) written together with Heinz Sauermann and published in 1959. When we began to do experimental economics at Frankfurt, such a field had not yet existed. My attempts to learn some psychology while I studied mathematics had made me acquainted to experimental techniques. I had listened to lectures of the gestalt psychologist Edwin Rausch, who was a careful experimenter, and I had participated in psychological experiments as a subject. Therefore it seemed natural to me to try an experimental approach to oligopoly.[7]

In the spirit of Skinner, game theorists such as Frederick Mosteller conducted experiments on rats to determine whether the animals could identify the choice with the highest probability of reward. Mosteller and Nogee's (1951) experiment in which subjects faced real consequences and received real money for their performance is particularly noteworthy.

Other experiments were conducted to examine the robustness of the assumptions of expected utility theory. In 1948, Preston and Baratta conducted an experiment to examine risk preferences in which participants bid in an auction to win the opportunity of placing a bet. Maurice Allais, a pioneer in behavioral economics, presented a decision problem known as the Allais paradox (1953), which showed inconsistency in choice with the independence assumption in expected utility theory. In terms of experimental methods, Allais was also a pioneer in the method of comparing different risky lotteries to examine risk behavior.

In 1954, William Estes, a psychologist and student of B. F. Skinner, conducted experiments to test mathematical learning models. His experiments used a performance-based payoff structure in which subjects were rewarded based on their probability of making a correct prediction (Estes 1955). In the late 1950s and early 1960s, Jacob Marschak conducted a significant series of experiments on stochastic decision-making (Davidson and Marschak 1959).

In 1953, Harold Kuhn (1953) first began analyzing games using an extensive form structure and introduced the notion of a subgame, a concept that Reinhard Selten (1965) extended to subgame perfect equilibrium, one of the first refinements of a Nash equilibrium.

In 1957, Duncan Luce and Howard Raiffa published the first "modern" textbook on game theory entitled *Games and Decisions: Introduction and Critical Survey*. Also important was the development of a body of "cross-over" books that showed the interdisciplinary nature of game theory.

During this time, the psychologist Sidney Siegel and the economist Lawrence Fouraker (Fouraker and Siegel 1977) began conducting experiments on bilateral bargaining. The methodology they used is reflected in the political economy experiments we recognize today, in which treatments vary the information to which subjects have access. Hence, the beginnings of experimental tests of game theory were developed by both economists and psychologists. It is widely believed that Siegel's untimely death delayed a "promising interdisciplinary venture in experimental gaming" (Dimand and Dimand 2005, 14).

Further development of game theory in the late sixties involved refinements to the Nash equilibrium. Prior to 1967, game theory consisted of full-information games. However, others had been thinking about this idea, with the most notable being Martin Shubik, who in 1952 wrote "Information, Theories of Competition, and the Theory of Games," a paper that was interested in how information exchanged between two players impacts their behavior. In 1967–68, John Harsanyi published three seminal papers that allowed models to incorporate incomplete information. In 1975, Selten created a refinement of the Nash equilibrium for incomplete information games using a Bayesian information structure called a perfect Bayesian equilibrium. In 1994, John Nash, John Harsanyi, and Reinhard Selten won the Nobel Prize in economics for their contributions to theoretical game theory. In 2009, Elinor Ostrom, a professor at Indiana University, also won the Nobel Prize, partly as a result of her application of game theory and experimental methods.

8. SOCIAL CHOICE EXPERIMENTS

Social choice theory examines deterministic voting systems in which a set of voters have defined preferences over a set of (usually) two or three alternatives. Some institution then selects a single winner (or several winners, depending on the institution). The question is, what type of properties in terms of fairness and equity of outcomes should this institution have? For example, let us assume that preferences are held fixed across two institutions, and that the only variable that varies between the two is that in one institution a single person wins, and in another two people win. If winners are given some payoff (since the institutions are exactly the same, all players will receive the same payoffs) that is divided differently among them, then this must impact issues of fairness and equity. The point is that different institutions produce different outcomes, so social choice theory is concerned with which institutions produce the most fair and equitable outcomes.

The roots of social choice theory date back to 1299, when Ramon Lull proposed that candidates in multi-candidate elections compete against one another two at a time. In 1785, the Marquis de Condorcet (probably the first formal political science theorist) wrote his *Essay on the Application of Analysis to the Probability of Majority Decisions*. In this essay, he laid out the following problem, referred to as the voter's paradox or the paradox of voting. Assume that three equally sized groups of voters have to decide among three alternatives, x, y, and z. Group 1 prefers x to y to z, Group 2 prefers y to z to x, and Group 3 prefers z to x to y. Now assume that the voting rule requires a pairwise comparison of alternatives. If x is paired against z, then z wins, since Groups 2 and 3 prefer z to x. Now pair z against y. Y wins, because Groups 1 and 2 both prefer y to z. Pairing y against x yields x as the eventual winner. This configuration of preferences yields what is called an intransitive social ordering, in which the winner of the election depends on the match-up with which voters are presented.

In 1951, Kenneth Arrow realized every graduate student's dreams of winning a Nobel Prize for his dissertation, *Social Choice and Individual Values*. Using the axiomatic method, he created a unified framework in which to evaluate the impact of rules on preferences. His approach applied to all conceivable rules that are based on preferences. Arrow's impossibility theorem, called the general possibility theorem or Arrow's paradox, states that when you consider three or more alternatives, you cannot construct a voting system that satisfies some set of fair criteria. This idea was important, because it pointed to the relevance of institutions and their ability to manipulate this choice function. That is, if there is no inherently fair voting system, then voting systems (or different institutions) have to be created that admit some sort of biases. The question then becomes, how fair are the biases? Or given a set of biased voting systems, which is the least biased?

Arrow's findings were bolstered by Robin Farquharson, who in 1956 showed the presence of strategic voting—or voting for a least preferred outcome to get a more preferred outcome—in deterministic voting rules. Allan Gibbard and Mark Satterthwaite subsequently proved that when there are three or more alternatives, all voting systems that pick a single winner are manipulable (Gibbard 1973; Satterthwaite 1975). In 1976, Richard D. McKelvey proved the chaos theorem, which specifies that if a Condorcet winner does not exist in a majority-rule institution, then any outcome within the space is possible. In short, these four results mean that given a set of preferences, all voting systems can be manipulated.

This result might be called the beginning of an institutional revolution. No longer is the behavior of preferences important as long as it is well behaved. Now it is the institution itself that is the focus of the study. For example, how does a majority-rule institution differ from an approval-voting institution, given the same set of voter preferences? We can now determine the bias of one system as opposed to another by holding preferences constant.

Buchanan and Tullock's *Calculus of Consent* (1962) and Riker's *Theory of Political Coalitions* (1962) were also important in the development of this institutional approach. Riker created a program at Rochester University that produced an army of political scientists intent on using this axiomatic approach to study politics in terms of institutional behavior.[8] Also important was Hayward Alker's *Mathematics and Politics* (1965), which helped pave the way for an axiomatic treatment of international relations problems. Other important works in this area include Thomas Schelling's *Strategy of Conflict* (1960) and the works of Richard McKelvey and Thomas Palfrey (1995; 1998), which offered a way to introduce uncertainty in equilibrium analysis in both extensive and normal form games.

Another important development was the publication of Duncan Black's *On the Rationale of Group Decision Making* (1948), which adopted some of Hotelling's work to prove that when you consider groups on a single dimension, equilibrium tends to form in the middle of the groups or, more specifically, at the median group's idea point. Black's work allowed the abstract dimensions and vector spaces with which game theorists had been working. These dimensions could represent policy positions such as liberal or conservative, or almost any other variable. Hence, we can understand a committee situation as members bargaining over alternatives in a one- or two-dimensional policy space. By bargaining, we mean moving one member's alternative on a dimension closer to or farther away from another member's ideal point on the same dimension. Therefore, in an experiment, a condition of stability can be established by controlling preferences, allowing various institutions to be studied.

In 1978, Morris Fiorina and Charles Plott established an experimental methodology to examine this structure (with their findings representing the first laboratory experiment published in a major political science journal). Gary Miller (2011, 354) describes their experimental method:

> They used students as subjects, presenting them with two-dimensional sets of possible decisions "outcomes"—the crucial dependent variable. The dimensions were presented in an abstract way intended to render them neutral of policy or personal preferences.
>
> The two dimensions were salient only for their financial compensation; the students saw payoff charts showing concentric circles around their highest-paying "ideal point." One student might receive a higher payoff in the upper right-hand corner, whereas others would prefer other areas in the space. The students were quite motivated by the payoffs, a fact that gave the experimenters control over the key independent variable—preferences.

This framework was adopted by Richard McKelvey and Peter Ordeshook (1985) to test election questions posed by Anthony Downs in his seminal book, *An Economic Theory of Democracy* (1957), in which he presented the median voter theorem. Melvin Hinich and Otto Davis (Davis and Hinich 1966) began the field of multidimensional spatial theory of elections. Hinich's test of the theory using estimates of spatial maps of candidates was a pioneer in the use of observational data to test formal models (Enelow and Hinich, 1984). Finally, in the mid 1980s the use of computer network technology driven by the economic auction experiments of Charles Holt, Charles Plott and Vernon Smith (see Plott, 1991) propelled computer network based experiments to examine committee institutions (developed at Rice University by Rick Wilson, see Wilson, 1986) and election institutions (developed at the University of Texas at Austin by Kenneth Collier, Peter Ordeshook, Richard McKelvey, and Kenneth Williams, see Collier et al. 1987, 1989).

9. CONCLUSIONS

One of the goals of this history was to show how the interaction of the axiomatic approach and probability theory provides an objective way to study social issues. The evolution we see today is natural. The axiomatic approach advocates starting from scratch and building a model from the ground up. Such an approach demands a framework, and I like to think of the researchers presented in this history as the suppliers of this rich framework in which to study human behavior.

Behavioral game theory is the second wave in this evolution. It is concerned with building upon this framework by adding psychological elements that might help us to better explain decision-making.

NOTES

1. Anyone with more than casual interest in the history of game theory should consult Leonard (2010).
2. This letter appears in Fermat (1800, 288–314).
3. R. A. Fisher (1934) also proved the minimax theorem independently for the case in which players have pure strategies.
4. As a side note, while doing graduate work at Princeton, John Nash created a game that colleagues called Nash or John (because it was played on the bathroom floor). This game was later named Hex when it was marketed by Parker Brothers.
5. Each player has an allocated color, red or blue. Players take turns placing a stone of their color on a single cell within the overall playing board, with the first player forbidden to move into the center. The goal is to form a connected path of stones linking the opposing sides of the board marked with your colors before your opponent connects his sides in a similar fashion. The first player to complete his connection wins the game. John Nash proved that the game can never end in a tie, and that the only way to prevent your opponent from forming a connecting path is to form a path yourself.
6. Flood's advisor Albert Tucker devised the prisoner's dilemma story.
7. See http://nobelprize.org/nobel_prizes/economics/laureates/1994/selten-autobio.html.
8. I am a descendent of this school and tradition through my two major advisors, Peter Ordeshook and Melvin Hinich.

SINCERE VS. STRATEGIC VOTING IN AGENDA GAMES

Consider a voting game with three voters who have ranked preferences as illustrated in Table A2.

In this case, there are three voters who have preferences over three alternatives x, y, and z. Voter 1 prefers x to y to z; Voter 2 prefers z to x to y; and Voter 3 prefers y to z to x. First let us consider the case in which voters vote among the alternatives using simple majority rule, in which the alternative with the most votes wins. Let us also assume that voters vote sincerely, voting for their most preferred alternative. In this election, if each voter votes for the preference he or she has ranked first, then the election outcome will be a tie, since Voter 1 will vote for x, Voter 2 will vote for z, and Voter 3 will vote for y. But let us consider an alternative voting procedure: pairwise voting, in which pairs of alternatives are voted on two at a time. Using pairwise voting, we can match the alternatives x and y. In this case, in the pairing of x and y, x wins, since Voters 1 and 2 prefer x to y; in the pairing of y and z, y wins, since Voters 1 and 3 prefer y to z; and finally, in the pairing of z and x, z wins, since Voters 2 and 3

TABLE A2 THREE VOTERS WITH PREFERENCES OVER THREE ALTERNATIVES

		Voters		
		1	2	3
	First	x	z	y
Preference Order	Second	y	x	z
	Third	z	y	x

TABLE A3 X IS A CONDORCET WINNER

		Voters		
		1	2	3
	First	x	x	y
Preference Order	Second	y	z	z
	Third	z	y	x

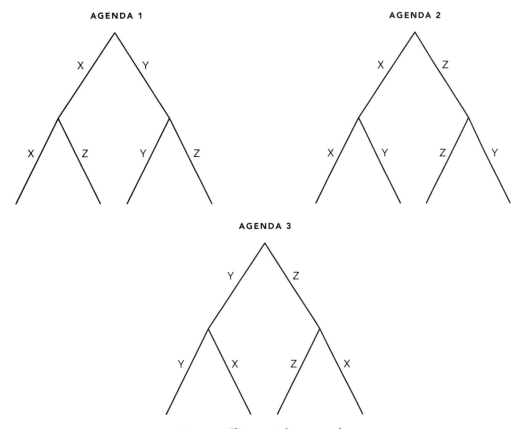

FIGURE A1 Three amendment agendas

prefer z to y. This is the famous example of the voting paradox, or the Condorcet paradox (Condorcet 1785). This preference ordering violates our assumption of transitivity, since there is no Condorcet winner that beats all other alternatives in a pairwise comparison. For example, if Player 2's preferences were switched so that her preferred alternative were x, as in Table A3, then x would be a Condorcet winner, since it would beat all others in a pairwise vote.

Now I would like to introduce the concept of amendment agendas, which are displayed in Figure A1. While they might look like the extensive form games that we have studied, they are not games, because they only specify the order of voting. The first agenda pits x against y, and the winner is then pitted against z (xyz); the second agenda pits x against z and the winner against y (xzy); and the last agenda pits y against z and the winner against x (yzx).

Using the preferences outlined in Table A3, and considering the case in which all voters vote sincerely, in the initial contest between x and y in the first agenda, Voters 1 and 2 prefer x to y, so x wins and we are on the left side of the agenda. The alternative x is then paired against z, which wins, since Voters 2 and 3 prefer z to x. In the second agenda y wins, and in the third agenda x wins. While this example is simple, it does illustrate the power of agendas and how the order in which players vote on alternatives can impact outcomes. Note that sincere voting is not optimal or subgame perfect, because voters in the game can achieve a better outcome. The reason for this is the assumption that voters who vote sincerely always vote for their most preferred alternative. Notice that z wins in the first agenda, but this outcome is Voter 1's last choice. This voter could do better by voting strategically. Strategic voting is voting for a least preferred alternative in order to achieve a more preferred outcome. If Player 1 were to use strategic voting on the first vote between x and y, instead of voting for x she should vote for y, giving y enough votes to beat z. To solve the strategic outcome for these types of games, we use our familiar method of backward induction. In the first agenda we pair x and z, and z wins; and we pair y and z, and y wins. The strategic outcome for the second agenda is x, and for the third agenda it is z. This method of backward induction, which takes into account sophisticated behavior, provides our subgame perfect equilibrium, in which no player has an action that can improve his or her payoff.

Now let us examine a strategic voting problem related to a coordination problem proposed by Myerson and Weber (1993) and first experimentally tested by Forsythe et al. (1993), called the Condorcet losers problem. Table A4 presents the payoffs for the experiment.

In this example, notice that if all voters voted sincerely in a three-way plurality election (in which the alternative who receives the most votes wins), then alternative C would win, since six voters who are Type C will vote for C, and Type A and B voters will vote for their associated type, respectively, giving each alternative four votes each. However, in a pairwise comparison, C will be a Condorcet loser, since he will lose in a two-way contest against the other two candidates. This payoff matrix presents a problem in the sense that Type A and Type B voters can beat C by strategic voting, but which type of voter should strategically vote for which candidate? Should Type A voters strategically switch and vote for B, or should Type B voters strategically switch and vote for Type A? This becomes a coordination problem in which Type A and Type B are both better off switching their votes. In experiments in which voters are just given the payoffs described and asked to vote for the three alternatives, Type A and Type B voters fail to coordinate and C wins 87.5 percent of the time.

What type of coordination devices can be implemented to achieve subgame perfect equilibrium? Forsythe et al.'s (1993) experiment used polls as a coordination device. Polls were implemented by having subjects vote on a nonbinding poll prior to voting in the real election. In this treatment, the Condorcet loser won only 33 percent of the time, and when either A or B were leading in the poll, the Condorcet loser won only 16 percent of the time.

In another experiment using the same Condorcet losing profile, Forsythe et al. (1996) varied the voting rules, specifically plurality, approval voting, and the Borda rule.

TABLE A4   CONDORCET LOSERS VOTER PAYOFF SCHEDULE

Voter type	Election Winner			Total number of each type
	A	B	C	
1 (A)	$1.20	$0.90	$0.20	4
2 (B)	$0.90	$1.20	$0.20	4
3 (C)	$0.40	$0.40	$1.40	6

Under plurality rules, the alternative with the most votes wins the election. Under approval voting rules, subjects vote for as many of the alternatives as they wish, and under the Borda rule, voters allocate two votes to their two most preferred alternatives and one vote to their least preferred alternative. The results showed that the Condorcet loser won more elections under plurality voting rules than under approval voting rules or the Borda rule. This finding illustrates that not all voting rules are equal, in that some provide more efficient or socially desirable outcomes than others.

Gerber, Morton, and Rietz (1998) used a payoff function similar to the one used in the previous experiment to compare cumulative versus straight voting in multimember districts (i.e., three candidates compete for two seats). In this experiment, the three voter types had the following preferences: four Type 1 voters preferred A to B to C, four Type 2 voters preferred B to A to C, and six Type 3 (minority) voters only gained utility from C. The payoffs were similar to those described in the Condorcet loser experiment. In cumulative voting, voters could cast two votes for one alternative, while in straight voting, voters could cast one vote for up to two alternatives. Voters were also given the option to not vote, or abstain, in the election. The results revealed that the minority alternative won more elections, since cumulative voting allowed minority types to cumulate their votes on the minority alternative. These results show that different coordination devices such as polls and alternative electoral systems allow subgame perfect equilibrium outcomes.

PROBLEM SETS

1. Consider the following profile. Create two agendas (in which members vote sincerely), with A being the winner in one agenda and D being the winner in the other agenda.

Voter 1	Voter 2	Voter 3
A	B	C
D	A	B
C	D	A
B	C	D

2. Use the following profiles and solve using an amendment agenda (abcd). Solve sincerely and sophisticatedly.

Voter 1	Voter 2	Voter 3
A	B	D
C	A	C
B	D	B
D	C	A

3. The following is an example of an elimination agenda, so for example, if A beats B, then A wins and voting stops, but if B wins, B is paired against C, and if B wins again, voting stops. If C wins, C is paired against D for a final vote. Use the following preference profiles to determine the winner of this agenda.

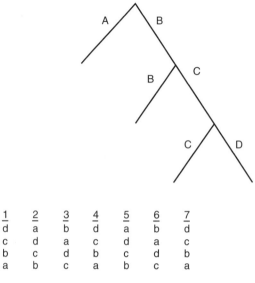

4   *Messy veto game.* In this agenda game, there are three players and four alternatives (a, b, c, d). Player 1 has the choice to veto one of the four alternatives, then Player 2 has the choice to veto one of the three remaining alternatives, and finally Player 3 has the option to veto one of the two remaining alternatives. The winning alternative is the one that is not vetoed. I call this a messy veto game because the agenda is rather cumbersome to draw. Solve the agenda sincerely and sophisticatedly using the following player preferences:

Player 1: a > b > c > d
Player 2: b > a > d > c
Player 3: d > b > a > c

The following figure is a partial representation of the agenda:

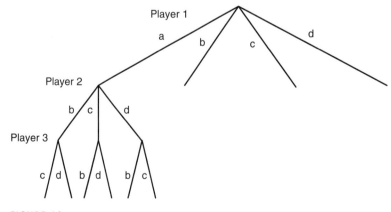

FIGURE A3

REFERENCES

Abrams, Robert. 1980. *Foundation of Political Analysis: An Introduction to the Theory of Collective Choice*. New York: Columbia University Press.

Ainslie, G., and N. Haslam. 1992. "Hyperbolic Discounting." In *Choice over Time*, ed. G. Loewenstein and J. Elster, 57–92. New York: Russell Sage.

Alker, Hayward. 1965. *Mathematics and Politics*. New York: Macmillan.

Allais, M. 1953. "Le Comportement de l'Homme Rationel devant le Risque: Critique des Postulates et Axiomes de l'Ecole Americaine." *Econometrica* 21:502–46.

Allsopp, L., and J. Hey. 2000. "Two Experiments to Test a Model of Herd Behavior." *Experimental Economics* 3:121–36.

Anderson, Lisa R., and Charles A. Holt. 1997. "Information Cascades in the Laboratory." *American Economic Review* 87 (5): 847–62.

Andreoni, James, and John H. Miller. 1993. "Rational Cooperation in the Finitely Repeated Prisoner's Dilemma: Experimental Evidence." *Economic Journal* 103:570–85.

Aranson, Peter H. 1981. *American Government: Strategy and Choice*. Cambridge, MA: Winthrop.

Ariely, Dan. 2008. *Predictably Irrational*. New York: Harper Collins.

Arrow, Kenneth 1951. *Social Choice and Individual Values*. New Haven, CT: Yale University Press.

Arthur, Brian W. 1994. "Inductive Reasoning and Bounded Rationality." *American Economic Review* 84:406–11.

Associated Press. 1991. "MIT Research Isn't All Quarks and Quasars, Sometimes It's Baseballs." *Ludington Daily News*, December 30. news.google.com/newspapers?nid=110&dat=19911230&id=qfZPAAAAIBAJ &sjid=SFUDAAAAIBAJ&pg=6724,6734553.

Austen-Smith, David. 1990. "Information Transmission in Debate." *American Journal of Political Science* 34 (1): 124–52.

Axelrod, Robert. 1984. *The Evolution of Cooperation*. New York: Basic.

Banerjee, A. 1992. "A Simple Model of Herd Behavior." *Quarterly Journal of Economics* 107:797–817.

Basmann, Robert L., Raymond C. Battalio, Leonard Green, John H. Kagel, W. R. Klemm, and Howard Rachlin. 1975. "Experimental Studies of Consumer Demand Behavior using Laboratory Animals." *Economic Inquiry* 13:22–38.

Bayes, Thomas. 1958. "An Essay towards Solving a Problem in the Doctrine of Chances." *Biometrika* 45:296–315.

Bell, D. E. 1982. "Regret in Decision-Making under Uncertainty." *Operations Research* 30:961–81.

Bennett, D. P. 2011. "Unbound or Distant Planetary Mass Population Detected by Gravitational Microlensing." *Nature* 473:349–52.

Berg, Joyce E., John Dickhaut, and Kelvin McCabe. 1995. "Trust, Reciprocity, and Social History." *Games and Economic Behavior* 10:122–42.

Bernoulli, Daniel. 1738. "Specimen theoriae novae do mensura sortis" ["Exposition of a New Theory on the Measurement of Risk"]. *Commentarii Academiae Scientiarum Imperialis Petropolitanae* 5:175–92.

Bikhchandani, Sushil, David Hirshleifer, and Ivo Welch. 1992. "A Theory of Fads, Fashion, Custom, and Cultural Change as Information Cascades." *Journal of Political Economy* 100:992–1026.

Binmore, Kenneth, Joe Swierzbinski, and Chris Proulx. 2001. "Does Minimax Work? An Experimental Study." *Economic Journal* 111:445–64.

Black, Duncan. 1948. "On the Rationale of Group Decision Making." *Journal of Political Economy* 56:23–34.

———. 1958. *A Theory of Committees and Elections*. Cambridge, UK: Cambridge University Press.

Bolton, Gary E., and Axel Ockenfels. 2000. "ERC: A Theory of Equity, Reciprocity, and Competition." *American Economic Review* 81:1096–1136.

Borel, Émile. 1938. *Traite du Calcul des Probabilites et de ses Applications*, Tome IV, Fasc. 2, Applications des Jeux de Hasard. Paris: Gauthier-Villars.

Bower, Amy S., M. Susan Lozier, Stefan F. Gary, and Claus W. Böning. 2009. "Interior Pathways of the North Atlantic Meridional Overturning Circulation." *Nature* 459:243–47.

Brams, Steven J., and D. Marc Kilgour. 1988. *Game Theory and National Security*. New York: Basil Blackwell.

Brown, James N., and Robert W. Rosenthal. 1990. "A Re-examination of O'Neill's Game Experiment." *Econometrica* 38:1065–81.

Buchanan, James, and Gordon Tullock. 1962. *The Calculus of Consent: Logical Foundations of Constitutional Democracy*. Indianapolis, IN: Liberty Fund, Inc.

Camerer, Colin F. 2003. *Behavioral Game Theory*. Princeton, NJ: Princeton University Press.

Camerer, Colin F., Teck-Hua Ho, and Juin-Kuan Chong. 2002. "Sophisticated Experience-Weighted Attractive Learning and Strategic Teaching in Repeated Games." *Journal of Economic Theory* 104 (1): 137–88.

Camerer, Colin, George Loewenstein, and M. Rabin. 2003. *Advances in Behavioral Economics*. Princeton, NJ: Princeton University Press.

Camerer, Colin F., and Richard H. Thaler. 1995. "Ultimatums, Dictators and Manners." *Journal of Economic Perspectives* 9:209–19

CBS News. 2009. "Casino Mogul Steve Winn's Midas Touch." *CBS News*, July 16. http://www.cbsnews.com/stories/2009/04/10/60minutes/main4935567.shtml.

Çelen, B., and S. Kariv. 2004. "Observational Learning under Imperfect Information." *Games and Economic Behavior* 94:484–98.

Chamberlin, Edward. 1933. *The Theory of Monopolistic Competition*. Cambridge, MA: Harvard University Press.

Charron, William. 2000. "Greeks and Games: The Ancient Forerunners of Mathematical Game Theory." *Forum for Social Economics* 29 (2): 1–32.

Chew, S. H., and K. R. MacCrimmon. 1979. "Alpha-Nu Choice Theory: A Generalization of Expected Utility Theory." Working Paper No. 669, University of British Columbia, Vancouver.

Cho, In-Koo, and David M. Kreps. 1987. "Signaling Games and Stable Equilibrium." *Quarterly Journal of Economics* 102:180–221.

Chong, Dennis, and James Druckman. 2007. "Framing Theory." *Annual Review of Political Science* 10:103–26.

Chou, Eileen, Margaret McConnell, Rosemarie Nagel, and Charles Plott. 2007. "The Control of Game Form Recognition in Experiments: Understanding Dominant Strategies Failures in a Simple Two Person 'Guessing' Game." Social Science Working Paper 1274, California Institute of Technology, Pasadena, CA.

Colman, Andrew. 1982. *Game Theory and Experimental Games: The Study of Strategic Interaction.* Oxford: Pergamon Press.

Collier, Kenneth, Richard McKelvey, Peter Ordeshook, and Kenneth Williams. 1987. "Retrospective Voting —An Experimental Study." *Public Choice* 53 (2): 101–310.

Collier, Kenneth, Peter Ordeshook, and Kenneth Williams. 1989. "The Rationally Uninformed Electorate: Some Experimental Results." *Public Choice* 60 (1): 3–29.

Condorcet, Marquis de. 1785. *Essai sur l'Application de l'Analyse à la Probabilité des Décisions Rendues à la Pluralité des Voix* (*Essay on the Application of Analysis to the Probability of Majority Decisions*). Paris: L'Imprimerie Royale.

Cook, Richard, Geoffrey Bird, Gabriele Lünser, Steffen Huck, and Cecilia Heyes. 2011. "Automatic Imitation in a Strategic Context: Players of Rock-Paper-Scissors Imitate Opponents' Gestures." *Proceedings of the Royal Society B* 279 (1729): 780–86.

Cook, Thomas D., and Donald T. Campbell. 1979. *Quasi-experimentation: Design and Analysis for Field Settings.* Boston, MA: Houghton Mifflin.

Cooper, Russell, Douglas V. DeJong, Robert Forsythe, and Thomas W. Ross. 1992. "Communication in Coordination Games." *Quarterly Journal of Economics* 107 (2): 739–71.

———. 1994. "Alternative Institutions for Resolving Coordination Problems: Experimental Evidence on Forward Induction and Preplay Communication." In *Problems of Coordination in Economic Activity*, ed. J. Friedman, 129–46. Dordrecht, The Netherlands: Kluwer Academic.

Dana, Jason D., Dalian M. Cain, and Robin M. Dawes. 2006. "What You Don't Know Won't Hurt Me: Costly (but Quiet) Exit in a Dictator Game." *Organizational Behavior and Human Decision Processes* 100:193–201.

Dasgupta, Sugato, Kirk A. Randazzo, Reginald S. Sheehan, and Kenneth C. Williams. 2008. "Coordinated Voting in Sequential and Simultaneous Elections: Some Experimental Results." *Experimental Economics* 11:315–35.

Davidson, Donald, and Jacob Marschak. 1959. "Experimental Tests of a Stochastic Decision Theory." In *Measurement: Definitions and Theories*, ed. C. West Churchman and Philburn Ratoosh, 233–70. New York: Wiley.

Davis, Douglas D., and Charles A. Holt. 1993. *Experimental Economics.* Princeton, NJ: Princeton University Press.

Davis, Morton. 1982. *Game Theory: A Nontechnical Introduction.* New York: Basic.

Davis, Otto, and Melvin Hinich. 1966. "A Mathematical Model of Policy Formulation in a Democratic Society." *Public Choice* 5:59–72.

DeQuervain, Dominique, Urs Fischbacher, Valerie Treyer, Melanie Schellhammer, Ulich Schnyder, Alfred Buck, and Ernst Fehr. 2004. "The Neural Basis of Altruistic Punishment." *Science* 305 (5688): 1254–58.

Deutsch, Morton, and R. M. Krauss. 1960. "The Effect of Threat on Interpersonal Bargaining." *Journal of Abnormal and Social Psychology* 61:181–89.

Deutsch, Morton, and Roy J. Lewicki. 1970. " 'Locking in' Effects during a Game of Chicken." *Journal of Conflict Resolution* 14:367–78.

Dimand, Robert W. 2005. "Experimental Economic Games: The Early Years." In *The Experiment in the History of Economics*, ed. Philippe Fontaine and Robert Leonard, 4–20. London: Routledge.

Dimand, Mary Ann, and Robert W. Dimand. 2005. *The History of Game Theory*, Vol. 1. London: Routledge.

Downs, Anthony. 1957. *An Economic Theory of Democracy*. New York: Harper and Row.

Druckman, James. 2001. "Evaluating Framing Effects." *Journal of Economic Psychology* 22:91–101.

Dufwenberg, Martin, Ramya Sundaram, and David Butler. 2008. "Epiphany in the Game of 21." Working paper, University of Arizona, Tempe, AZ.

Eckel, Catherine C., and Rick K. Wilson. 2007. "Social Learning in Coordination Games: Does Status Matter?" *Experimental Economics* 10:317–29.

Eisenberg, Nancy, and Paul H. Mussen. 1989. *The Roots of Prosocial Behavior in Children*. Cambridge, UK: Cambridge University Press.

Ellsberg, Daniel. 1961. "Risk, Ambiguity and the Savage Axioms." *Economic Journal* 64:453–82.

Elster, Jon. 1998. "Emotions and Economic Theory." *Journal of Economic Literature* 36 (1): 47–74.

Enelow, James, and Melvin Hinich. 1984. *The Spatial Theory of Voting: An Introduction*. Cambridge, UK: Cambridge University Press.

Estes, W. K. 1955. "Statistical Theory of Distributional Phenomena in Learning" *Psychological Review* 62 (5): 369–77.

Farquharson, Robin. 1956. "Straightforwardness in Voting Procedures." Oxford Economic Papers, New Series 8 (1):80–89.

Fehr, Ernst, and Urs Fischbacher. 2003. "The Nature of Human Altruism." *Nature* 425 (6960): 785–91.

Fehr, Ernst, and Simon Gächter. 2000. "Altruistic Punishment in Humans." *Nature* 415 (6868):137–40.

Fehr, Ernst, and Klaus M. Schmidt. 1999. "A Theory of Fairness, Competition, and Cooperation." *Quarterly Journal of Economics* 114 (3): 817–68.

Fermat, Pierre de. 1800. *Œuvres de Fermat*, ed. Paul Tannery and Charles Henry. Paris: Ministere d'Instruction.

Fiorina, Morris P., and Charles R. Plott. 1978. "Committee Decisions under Majority Rule: An Experimental Study." *American Political Science Review* 72:575–98.

Fisher, R. A. 1934. "Randomisation, and an Old Enigma of Card Play." *Mathematical Gazette* 18:294–97.

Flood, Merrill M. 1958. "Some Experimental Games." *Management Science* 5:5–26.

Forsythe, Robert, Roger Myerson, Thomas Rietz, and Robert Weber. 1993. "An Experimental Study on Coordination in Multi-Candidate Elections: The Importance of Polls and Election Histories." *Social Choice and Welfare* 10:223–47.

———. 1996. "An Experimental Study of Voting Rules and Polls in a Three-Way Election." *International Journal of Game Theory* 25:355–83.

Fouraker, Lawrence, and Sidney Siegel. 1977. *Bargaining and Group Decision Making: Experiments in Bilateral Monopoly*. Westport, CT: Greenwood.

Fox, Craig, and Amos Tversky. 1995. "Ambiguity Aversion and Comparative Ignorance." *Quarterly Journal of Economics* 110 (3): 585–603.

Friedman, Daniel, and Shyam Sunder. 1994. *Experimental Methods: A Primer for Economists*. Cambridge, UK: Cambridge University Press.

Fudenberg, Drew, and Jean Tirole. 1991. *Game Theory*. Cambridge, MA: MIT Press.

Geanakoplos, John, David Pearce, and Ennio Stacchetti. 1989. "Psychological Game and Sequential Rationality." *Games and Economic Behavior* 1:60–79.

Gerber, Alan and Donald P. Green. 2000. "The Effects of Canvassing, Telephone Calls, and Direct Mail on Voter Turnout: A Field Experiment." *American Political Science Review* 94:653–63.

Gerber, Elisabeth, Rebecca Morton, and Thomas Rietz. 1988. "Minority Representation in Multimember Districts." *American Political Science Review* 92:127–44.

Gibbard, Allan. 1973. "Manipulation of Voting Schemes: A General Result." *Econometrica* 41 (4): 587–601.

Gilligan, Thomas, and Keith Krehbiel. 1987. "Collective Decision Making and Standing Committees: An Informational Rationale for Restrictive Amendment Procedures." *Journal of Law, Economics, and Organizations* 3:287–335.

Gintis, Herbert. 2009. *The Bounds of Reason*. Princeton, NJ: Princeton University Press.

Gneezy, Uri. 2005. "Deception: The Role of Consequences." *American Economic Review* 95:384–94.

Gneezy, Uri., E. Haruvy, and H. Yafe. 2004. "The Inefficiency of Splitting the Bill: A Lesson in Institution Design." *Economic Journal* 114:265–80.

Gneezy, Uri, Aldo Rustichini, and Alexander Vostroknutov. 2007. "I'll Cross That Bridge When I Come to It: Backward Induction as a Cognitive Process." Working Paper, University of California–San Diego.

Goeree, J. K., C. Holt, and T. Palfrey. 2008. "Quantal Response Equilibria." In *New Palgrave Dictionary of Economics*, 2nd ed, ed. Steven N. Durlauf and Lawrence E. Blume. New York: Palgrave Macmillan.

Green, Leonard, Howard Rachlin, and John Hanson. 1983. "Matching and Maximizing with Concurrent Ratio-Interval Schedules." *Journal of the Experimental Analysis of Behavior* 40:217–24.

Güth, Werner, Rolf Schmittberger, and Bernd Schwarze. 1982. "An Experimental Analysis of Ultimatum Bargaining." *Journal of Economic Behavior and Organization* 3:367–88.

Hardin, R. 1982. *Collective Action*. Baltimore, MD: Johns Hopkins University Press.

Harmon, Katherine. 2009. "Bizarre Planet Found to Orbit Backward." *Scientific American*, August 13. http://www.scientificamerican.com/blog/60-second-science/post.cfm?id=bizarre-planet-found-to-orbit-backw-2009-08-13.

Harsanyi, John. 1967–68. "Games with Incomplete Information Played by Bayesian Players." *Management Science* 14:159–82, 320–34, 486–502.

Hawking, Stephen. 1988. *A Brief History of Time*. New York: Bantam

Haywood, Oliver G., Jr. 1954. "Military Decision and Game Theory." *Operations Research* 2 (4): 365–85.

Heisenberg, Warner. 1927, "Über den anschaulichen Inhalt der quantentheoretischen Kinematik und Mechanik." *Zeitschrift für Physik* 43 (3–4): 172–98.

Helbing, Dirk, Martin Schönhof, Hans-Ulrich Stark, and Janusz A. Holyst. 2005. "How Individuals Learn to Take Turns: Emergence of Alternating Cooperation in a Congestion Game and the Prisoner's Dilemma." *Advances in Complex Systems* 8:87–116.

Henrich, Joseph. 2000. "Does Culture Matter in Economic Behavior? Ultimatum Game Bargaining among the Machiguenga of the Peruvian Amazon." *American Economic Review* 90:973–79.

Henrich, Joseph, Robert Boyd, Samuel Bowles, Colin Camerer, Ernst Fehr, Herbert Gintis, and Richard McElreath. 2001. "In Search of Homo Economicus: Behavioral Experiments in 15 Small-Scale Societies." *American Economic Review* 91:73–78.

Hertwig, Ralph, and Andreas Ortmann. 2001. "Experimental Practices in Economics: A Methodological Challenge for Psychologists?" *Behavioral and Brain Sciences* 24:383–451.

Heyman, Gene M., and R. J. Herrnstein. 1986. "More on Concurrent Interval-Ratio Schedules: A Replication and Review." *Journal of the Experimental Analysis of Behavior* 46:331–51.

Hilbert, David. 1899. *Grundlagen der Geometrie*. Leipzig: Teubner.

Hinich, Melvin J., and Michael C. Munger. 1997. *Analytical Politics*. Cambridge, UK: Cambridge University Press.

Hirshleifer, Jack. 1989. "Conflict and Settlement." In *Game Theory*, ed. John Eatwell, Murray Milgate, and Peter Newman, 86–94. New York: Norton.

Hogarth, Robin, and Melvin W. Reder. 1986. "Introduction: Perspectives from Economics and Psychology." In *Rational Choice: The Contrast between Economics and Psychology*, ed. Robin M. Hogarth and Melvin W. Reder, 1–23. Chicago: University of Chicago Press.

Hotelling, Harold. 1929. "Stability and Competition." *Economic Journal* 39 (153): 41–57.

Hung, Angela, and Charles Plott. 2001. "Information Cascades: Replication and an Extension to Majority Rule and Conformity-Rewarding Institutions." *American Economic Review* 91:1508–20.

Hurd, Peter L. 1995. "Communication in Discrete Action-Response Games." *Journal of Theoretical Biology* 174:217–22.

Hy, Ronan Le, Anothony Arrigoni, Pierre Bessière, and Olivier Lebeltel. 2004. "Teaching Bayesian Behaviours to Video Game Characters." *Robotics and Autonomous Systems* 47:177–85.

Innocenti, Alessandro, and Patrizia Sbriglia, eds. 2008. *Games, Rationality and Behaviour Eassays on Behavioural Game Theory.* New York: Palgrave Macmillan.

Isaac, R. M., J. M. Walker, and A. Williams. 1994. "Group Size and the Voluntary Provision of Public Goods—Experimental Evidence Utilizing Large Groups." *Journal of Public Economics* 54:1–36.

Johnson, Eric J., Colin Camerer, Sankar Sen, and Talia Rymon. 2002. "Detecting Failures of Backward Induction: Monitoring Information Search in Sequential Bargaining Experiments." *Journal of Economic Theory* 104:16–47.

Kagel, John H., Raymond C. Battalio, Howard Rachlin, Leonard Green, Robert L. Basmann, and W. R. Klemm. 1975. "Experimental Studies of Consumer Demand Behavior using Laboratory Animals." *Economic Inquiry* 13 (1): 22–28.

Kahneman, Daniel, Jack L. Knetsch, and Richard Thaler. 1986. "Fairness and the Assumption of Economics." In *Rational Choice: The Contrast between Economics and Psychology*, ed. Robin M. Hogarth and Melvin W. Reder, 101–16. Chicago: University of Chicago Press.

Kahneman, Daniel, and Amos Tversky. 1979. "Prospect Theory: An Analysis of Decision under Risk." *Econometrica* 47:263–91.

Kaku, Michio. 2009. *Physics of the Impossible.* London: Penguin.

Keynes, John Maynard. 1936. *The General Theory of Employment, Interest and Money.* New York: Harcourt Brace and Co.

Knight, Frank. 1921. *Risk, Uncertainty, and Profit.* Boston: Houghton Mifflin.

Kosfeld, Michael, Markus Heinrichs, Paul Zak, Urs Fishbacher, and Ernst Fehr. 2005. "Oxytocin Increases Trust in Humans." *Nature* 435 (7042): 673–76.

Kreps, David, and Robert Wilson. 1982. "Sequential Equilibria." *Econometrica* 50:863–94.

Kuhn, Harold. 1953. "Extensive Games and the Problem of Information." In *Contributions to the Theory of Games*, vol. 2, ed. Harold Kuhn and Albert Tucker. Annals of Mathematics Studies, No. 28. Princeton, NJ: Princeton University Press.

Lasker, Emanuel. 1947. *Lasker's Manual of Chess, with 308 Diagrams.* Philadelphia: David McKay.

Ledyard, John. 1995. "Public Goods: A Survey of Experimental Research" In *Handbook of Experimental Economics*, ed. J. Kagel and A. Roth, 111–94. Princeton, NJ: Princeton University Press.

Leonard, Robert. 2010. *Von Neumann, Morgenstern, and the Creation of Game Theory: From Chess to Social Science, 1900–1960.* Historical Perspectives on Modern Economics. New York: Cambridge University Press.

Levitt, Steven, and John List. 2007. "What Do Laboratory Experiments Measuring Social Preferences Reveal About the Real World?" *Journal of Economic Perspectives* 21 (2): 153–74.

Levitt, Steven, John List, and Sally Sadoff. 2009. "Checkmate: Exploring Backward Induction among Chess Players." National Bureau of Economic Research Working Paper No. 15610, Washington, DC.

Loewenstein, G., L. Thompson, and M. Bazerman. 1989. "Social Utility and Decision Making in Interpersonal Contexts." *Journal of Personality and Social Psychology* 57:426–41.

Loomes, Graham, and Robert Sugden. 1982. "Regret Theory: An Alternative Theory of Rational Choice under Uncertainty." *Economic Journal* 92:805–24.

Luce, Duncan L., and Howard Raiffa. 1957. *Games and Decisions: Introduction and Critical Survey.* New York: Wiley.

Lupia, Arthur, and Mathew D. McCubbins. 1998. *The Democratic Dilemma.* Cambridge, UK: Cambridge University Press.

Machina, Mark J. 1987. "Choice under Uncertainty: Problems Solved and Unsolved." *Journal of Economic Perspectives* 1:121–54.

Markowitz, Harry. 1952. "The Utility of Wealth." *Journal of Political Economy* 60:151–58.

Matthews, Steven. 1989. "Veto Threats: Rhetoric in a Bargaining Game." *Quarterly Journal of Economics* 104 (2): 347–69.

May, Kenneth O. 1954. "Intransitivity, Utility, and the Aggregation of Preference Patterns." *Econometrica* 22:1–13.

Maynard Smith, John, and David Harper. 2003. *Animal Signals.* Oxford: Oxford University Press.

Maynard Smith, John, and G. R. Price. 1973. "The Logic of Animal Conflicts." *Nature* 246:15–18.

McCarty, Nolan, and Adam Meirowitz. 2007. *Political Game Theory.* Cambridge, UK: Cambridge University Press.

McCornack, S. A., and M. R. Parks. 1986. "Deception Detection and Relationship Development: The Other Side of Trust." In *Communication Yearbook 9*, ed. M. L. McLaughlin, 377–89. Beverly Hills, CA: Sage.

McCulloch, Warren S. 1948. "A Recapitulation of the Theory, with a Forecast of Several Extensions." *Teleogical Mechanisms, Annals of the New York Academy of Science* 50:259–77.

McKelvey, Richard. 1976. "Intransitivities in Multidimensional Voting Models and Some Implications for Agenda Control." *Journal of Economic Theory* 12:472–82.

McKelvey, Richard, and Peter C. Ordeshook. 1985. "Elections with Limited Information: A Fulfilled Expectations Model using Contemporaneous Poll and Endorsement Data as Information Sources." *Journal of Economic Theory* 36:55–85.

McKelvey, Richard, and Thomas Palfrey. 1992. "An Experimental Study of the Centipede Game." *Econometrica* 60:803–36.

———. 1995. "Quantal Response Equilibria for Normal Form Games." *Games and Economic Behavior* 10:6–38.

———. 1998. "Quantal Response Equilibria for Extensive Form Games." *Experimental Economics* 1:9–41.

Mehta, Judith, Chris Starmer, and Robert Sugden. 1994. "Focal Points in Pure Coordination Games: An Experimental Investigation." *Theory and Decision* 36:163–85.

Milgram, Stanley. 1963. "Behavioral Study of Obedience." *Journal of Abnormal and Social Psychology* 67 (4): 371–78.

Miller, Gary. 2011. "Legislative Voting and Cycling." In *The Cambridge Handbook of Experimental Political Science*, ed. James Druckman, Donald P. Green, James H. Kuklinski, and Arthur Lupia, 353–68. New York: Cambridge University Press.

Morrow, James D. 1994. *Game Theory for Political Scientists.* Princeton, NJ: Princeton University Press.

Morton, Rebecca B. 2007. "Why the Centipede Game Is Important." In *Positive Changes in Political Science: The Legacy of Richard D. McKelvey's Most Influential Writings*, ed. John H. Aldrich, James E. Alt, and Arthur Lupia, 365–76. Ann Arbor: University of Michigan Press.

Morton, Rebecca B., and Kenneth C. Williams. 2010. *Experimental Political Science and the Study of Causality: From Nature to the Lab.* Cambridge: Cambridge University Press.

Morton, Rebecca B., and Kenneth C. Williams. 2010. *Experimental Political Science and the Study of Causality: From Nature to the Lab.* Cambridge: Cambridge University Press.

———. 2011. "Electoral Systems and Strategic Voting (Laboratory Election Experiments)." In *The Cambridge Handbook of Experimental Political Science*, ed. James N. Druckman, Donald P. Green, James H. Kuklinski, and Arthur Lupia, 369–83. New York: Cambridge University Press.

Mosteller, F., and P. Nogee. 1951. "An Experimental Measurement of Utility." *Journal of Political Economy* 59:371–404.

Moulin, Hervé. 1986. *Game Theory for the Social Sciences.* New York: New York University Press.

Myerson, Roger B. 2004. "Comments on 'Games with Incomplete Information Played by "Bayesian" Players, I–III': Harsanyi's Games with Incomplete Information." *Management Science* 50 (12): 1818–24.

Myerson, Robert B., and Robert J. Weber. 1993. "A Theory of Voting Equilibria." *American Political Science Review* 87:102–14.

Nagel, R. 1998. "A Survey of Experimental Beauty Contest Game: Bounded Rationality and Learning." In *Games and Human Behavior, Essays in Honor of Amnon Rapoport*, ed. D. Budescu, I. Erev, and R. Zwick, 105–42. Mahwah, NJ: Erlbaum.

Nash, John. 1950. "The Bargaining Problem." *Econometrica* 18:155–62.

Ochs, Jack. 1995. "Games with Unique Mixed Strategy Equilibrium: An Experimental Study." *Games and Economic Behavior* 10:202–17.

O'Neill, Barry. 1987. "Nonmetric Test of the Minimax Theory of Two-Person Zero-Sum Games." *Proceedings of the National Academy of Sciences* 84:2106–09

Ordeshook, Peter C. 1986. *Game Theory and Political Theory: An Introduction*. Cambridge, UK: Cambridge University Press.

Osborne, Martin J. 2004. *An Introduction to Game Theory*. Oxford: Oxford University Press.

Osborne, Martin J., and Ariel Rubinstein. 1994. *A Course in Game Theory*. Cambridge: Massachusetts Institute of Technology Press.

Ostrom, Elinor. 2000. "Collective Action and the Evolution of Social Norms." *Journal of Economic Perspectives* 14:137–58.

———. 2009. "Beyond Market and States: Polycentric Governance of Complex Economic Systems." Nobel Prize Lecture, delivered at the Workshop in Political Theory and Policy Analysis, Indiana University, December 8.

Palacios-Huerta, Ignacio. 2003. "Professionals Play Minimax." *Review of Economic Studies* 70:395–415.

Palacios-Huerta, Ignacio, and Oscar Volij. 2009. "Field Centipedes." *American Economic Review* 99 (4): 1619–35.

Palfrey, Thomas R. 2006. "Laboratory Experiments." In *The Oxford Handbook of Political Economy*, ed. Barry Weingast and Donald Wittman, 915–36. New York: Oxford University Press.

———. 2007. "McKelvey and Quantal Response Equilibrium." In *Positive Changes in Political Science: The Legacy of Richard D. McKelvey's Most Influential Writings*, ed. John H. Aldrich, James E. Alt, and Arthur Lupia, 425–40. Ann Arbor: University of Michigan Press.

Palmer, Betsy. 2006. "Voting and Quorum Procedures in the Senate." CRS Report for Congress, Doc. No. 96-452. Washington, DC: Congressional Research Service.

Plott, Charles R. 1976. "Axiomatic Social Choice Theory: An Interpretation and Overview." *American Journal of Political Science* 20:511–96.

———. 1990. "Will Economics Become an Experimental Science?" Social Science Working Paper 758, California Institute of Technology, Pasadena, CA.

———. 1991. *Market Institutions and Price Discovery*. Cheltenham, UK: Edward Elgar.

Poundstone, Williams. 1992. *Prisoner's Dilemma*. New York: Anchor.

Preston, M. G., and P. Baratta. 1948. "An Experimental Study of the Auction Value of an Uncertain Outcome." *American Journal of Psychology* 61:183–93.

Quiggin, J. 1982. "A Theory of Anticipated Utility." *Journal of Economic Behavior and Organization* 3:323–43.

Rabin, Matthew. 1993. "Incorporating Fairness into Game Theory and Economics." *American Economic Review* 83:1281–1302.

Rapoport, A., and A. Chammah. 1965. *Prisoner's Dilemma*. Ann Arbor: University of Michigan Press.

Rasmussen, Eric. 1989. *Games and Information*. Oxford: Blackwell.

Riker, William. 1962. *The Theory of Political Coalitions*. New Haven, CT: Yale University Press.

Rogers, James R. 2001. "Information and Judicial Review: A Signaling Game of Legislative-Judicial Interaction. *American Journal of Political Science* 45 (1): 84–99.

Rosenthal, R. 1981. "Games of Perfect Information, Predatory Pricing, and the Chain Store." *Journal of Economic Theory* 25:92–100.

Roth, Alvin E. 1995a. "Bargaining Experiments." In *The Handbook of Experimental Economics*, ed. J. Kagel and A. Roth, 253–348. Princeton, NJ: Princeton University Press.

———. 1995b. "Introduction to Experimental Economics." In *The Handbook of Experimental Economics*, ed. J. Kagel and A. Roth, 3–20. Princeton NJ: Princeton University Press.

Roth, Alvin E., and Michael Malouf. 1979. "Game-Theoretic Models and the Role of Information in Bargaining." *Psychological Review* 86:574–94.

Roth, Alvin E., and Marilda A. Oliveira Sotomayor. 1990. *Two-Sided Matching: A Study in Game-Theoretic Modeling and Analysis.* Cambridge, UK: Cambridge University Press.

Roth, Alvin E., Vesna Prasnikar, Masahiro Okuno-Fujiwara, and Shmuel Zamir. 1991. "Bargaining and Market Behavior in Jerusalem, Ljubljana, Pittsburgh and Tokyo: An Experimental Study." *American Economic Review* 81:1068–95.

Rousseau, Jean-Jacques. 1755. Discourse on the Origin of Inequality Among Men. Denis Diderot's Encyclopédie.

Russell, Bertrand. 1959. *Common Sense and Nuclear Welfare.* New York: Simon and Schuster.

Samilton, Tracy. 2011. "Ford and the Case of the Chinese Official's Hat." *Michigan Radio*, October 14. http://michiganradio.org/post/ford-and-case-chinese-officials-hat.

Sanfey, Alan G., James K. Rilling, Jessica A. Aronson, Leigh E. Nystrom, and Jonathan Cohen. 2003. "The Neural Basis of Economic Decision-Making in the Ultimatum Game." *Science* 300:1755–58.

Satterthwaite, Mark A. 1975. "Strategy-Proofness and Arrow's Conditions: Existence and Correspondence Theorems for Voting Procedures and Social Welfare Functions." *Journal of Economic Theory* 10:187–217.

Schelling, Thomas. 1960. *Strategy of Conflict.* Oxford: Oxford University Press.

Schuck-Paim, Cynthia, Lorena Pompilio, and Alex Kacelnik. 2004. "State-Dependent Decisions Cause Apparent Violations of Rationality in Animal Choice." *PLoS Biology* 2 (12): e402.

Schwartz, Barry. 2004. *The Paradox of Choice: Why More Is Less.* New York: Harper Collins.

Schwartz, Hugh. 1998. *Rationality Gone Awry? Decision Making Inconsistent with Economic and Financial Theory.* Westport, CT: Praeger.

Selten, Reinhard. 1965. "Spieltheoretische Behandlung eines Oligopolmodells mit Nachfrageträgheit." *Zeitschrift für die gesamte Staatswissenschaft* 12:301–24.

———. 1975. "Re-examination of the Perfectness Concept for Equilibrium Points in Extensive Games." *International Journal of Game Theory* 4:25–55.

Shachat, Jason M. 2002. "Mixed Strategy Play and the Minimax Hypothesis." *Journal of Economic Theory* 104:189–226.

Shadish, W. R., T. D. Cook, and D. T. Campbell. 2002. *Experimental and Quasi-Experimental Designs for Generalized Causal Inference.* Boston: Houghton-Mifflin.

Shubik, Martin. 1952. "Information, Theories of Competition, and the Theory of Games." *Journal of Political Economy* 60:142–50.

Simon, Herbert. 1965. "The Logic of Rational Decision." *British Journal for the Philosophy of Science* 16:169–86.

———. 1985. "A. Human Nature in Politics: The Dialogue of Psychology with Political Science." *American Political Science Review* 79:293–304.

Skinner, B. F. 1938. *The Behavior of Organisms: An Experimental Analysis.* Cambridge, MA: B. F. Skinner Foundation.

Slovic, Paul, and Sarah Lichtenstein. 1983. "Preference Reversal: A Broader Perspective." *American Economic Review* 73:596–605.

Smith, Vernon L. 1962. "An Experimental Study of Competitive Market Behavior." *Journal of Political Economy* 70:111–37.

———. 1976. "Experimental Economics: Induced Value Theory." *American Economic Review* 66:274–79.

Solnick, Sara. J., and Maurice E. Schweitzer. 1999. "The Influence of Physical Attractiveness and Gender on Ultimatum Game Decisions." *Organizational Behavior and Human Decision Processes* 79 (3): 199–215.

Spence, Michael. 1974. *Market Signaling: Informational Transfer in Hiring and Related Screening Processes.* Cambridge, MA: Harvard University Press.

Stein, Jeremy. 1989. "Cheap Talk and the Fed: A Theory of Imprecise Policy Announcements." *American Economic Review* 79 (1): 32–42.

Strom, Gerald S. 1990. *The Logic of Lawmaking*. Baltimore, MD: John Hopkins University Press.

Sugden, Robert. 1982. "On the Economics of Philanthropy." *Economic Journal* 92:341–50.

Thrall, R., C. H. Combs, and R. L. Davis, eds. 1954. *Decision Processes*. New York: John Wiley.

Thurstone, Louis Leon. 1931. "The Indifference Function." *Journal of Social Psychology* 2:139–67.

———. 1959. *The Measurement of Value*. Chicago: University of Chicago Press.

Tversky, Amos, and Daniel Kahneman. 1986a. "The Framing of Decisions and the Psychology of Choice." In *Rational Choice*, ed. Jon Elster, 123–41. New York: New York University Press.

———. 1986b. "Rational Choice and the Framing of Decisions." *Journal of Business* 59:251–78.

Van Damme, E. 2002. *Stability and Perfection of Nash Equilibria*. 2nd ed. Heidelberg: Springer Verlag.

Van Doren, Carl, ed. 1945. *Benjamin Franklin's Autobiography*. New York: Viking.

Van Huyck, John B., John Wildenthal, and Raymond C. Battalio. 1992. "Tactic Cooperation, Strategic Uncertainty, and Coordination Failure: Evidence from Repeated Dominance Solvable Games." *Games and Economic Behavior* 38:156–75.

von Neumann, John. 1928. "Zur Theories der Gesellschafsspiele." *Mathematische Annalen* 100:295–320.

von Neumann, John, and Oskar Morgenstern. 1944. *The Theory of Games and Economic Behavior*. Princeton, NJ: Princeton University Press.

Walker, Mark, and John Wooders. 2001. "Minimax Play at Wimbledon." *American Economic Review* 91:1521–38.

Wallis, W. Allen, and Milton Friedman. 1942. "The Empirical Derivation of Indifference Functions." In *Studies in Mathematical Economics and Econometrics in Memory of Henry Schultz*, ed. O. Lange, F. McIntyre, and T. O. Yntema, 175–89. Chicago: University of Chicago Press.

Walsh, John I. 2007. *Do States Play Signaling Games? Cooperation and Conflict*. Thousand Oaks, CA: Sage.

Watson, Joel. 2002. *Strategy: An Introduction to Game Theory*. New York: W. W. Norton and Company.

Webster, Murray, Jr., and Jane Sell. 2007. "Theory and Experimentation in the Social Sciences." In *The Sage Handbook of Social Science Methodology*, ed. William Outhwaite and Stephen P. Turner, 190–207. Thousand Oaks, CA: Sage.

Whitt, Sam, and Rick Wilson. 2007. "The Dictator Game, Fairness and Ethnicity in Postwar Bosnia." *American Journal of Political Science* 51:655–68.

Willinger, Marc, Christopher Lohmann, and Jean-Claude Usunier. 2003. "A Comparison of Trust and Reciprocity between France and Germany: Experimental Investigation Based on the Investment Game." *Journal of Economic Psychology* 24:447–66.

Wilson, Rick K. 2005. "Anger, Fairness and What's in the Brain." Paper presented at the Festschrift in Honor of Elinor Ostrom, Workshop in Political Theory and Policy Analysis, Bloomington, Indiana, November 22–23.

Wooders, John. 2010. "Does Experience Teach? Professionals and Minimax Play in the Lab." *Econometrica* 78:1143–54.

World RPS Society. 2009. "Introduction: Why Study RPS?" http://www.worldrps.com/index.php?option=com_content&task=view&id=13&Itemid=28.

Xiao, Erte, and Daniel Houser. 2005. "Emotion Expression in Human Punishment Behavior." *Proceedings of the National Academy of Science* 102 (20): 7398–401.

Zaatari, Darine, Brian G. Palestis, and Robert Trivers. 2009. "Fluctuating Asymmetry of Responders Affect Offers in the Ultimatum Game Oppositely According to Attractiveness or Need as Perceived by Proposers." *Ethology* 115:627–32.

Zamir, Shmuel. 2000. "Rationality and Emotions in Ultimatum Bargaining." Lecture presented at the Conférence des Annales, Paris, June 19. http://www.ma.huji.ac.il/~zamir/dp222.pdf.

Zermelo, E. 1913. "Über eine Anwendung der Mengenlehre auf die theorie des Schachspiel." *Proceedings of the Fifth International Congress of Mathematics* 2:501–04.

GLOSSARY

AMENDMENT AGENDA   a voting agenda that specifies the order in which alternatives are voted on

ANONYMITY   the concealment of a subject's identity

APPLIED MODEL   a model that is built to be empirically tested

APPROVAL VOTING   a system of voting in which voters choose as many alternatives as they wish

ARTIFICIAL ENVIRONMENT   an environment created or manufactured by a researcher

ASYMMETRIC INFORMATION   a game structure in which one player has private payoff-specific information to which other players are not privy

AUDIENCE   multiple receivers in a signaling game

AUTOMATIC IMITATION (OR REINFORCEMENT) LEARNING   a system of learning in which players mimic actions that have been successful for other players in the past

BACKWARD INDUCTION (ALSO CALLED REDUCTION OR ROLLBACK EQUILIBRIUM)   a method of solving an extensive form game that considers actions in a path from the bottom of the game tree to the top of the game tree

BASELINE COMPARISON   an experiment that is conducted as a basis for comparing or contrasting experimental results

BATTLE OF THE SEXES   a two-player normal form game that illustrates coordination problems

BAYESIAN ANALYSIS   a method of updating beliefs based on conditional probabilities with relevant payoff-specific information

BAYESIAN NASH EQUILIBRIUM   a Nash equilibrium in which the equilibrium takes sequential rationality and conditional probabilities into account

BEHAVIORAL GAME THEORY   the use of psychological variables to explain behavior within a game theory model

BEHAVIORAL STRATEGIES   a complete mapping of a player's strategies given another player's strategies

BELIEFS   a probability distribution that specifies a player's likely location in a game tree, or the beliefs players have about random variables within the game structure

BELIEF LEARNING   a learning model in which a player builds a model of his or her opponent's possible moves in order to learn to respond to his or her actions

BELIEFS OFF THE EQUILIBRIUM PATH   a player's beliefs when he or she has no move in equilibrium

BEST RESPONSE FUNCTION the set of probabilities that a player assigns to a strategy in response to beliefs about what another player will play

BEST RESPONSE STRATEGY a strategy from which a player has no incentive to deviate

BETWEEN-SUBJECT DESIGN an experimental design with at least two treatments in which subjects are only exposed to one of the treatments

BINARY LOTTERY EXPERIMENT experiments in which subject preferences are modeled as lotteries with probability 1 assigned to one strategy and probability 0 assigned to all other strategies

BORDA RULE a voting rule in which voters allocate two votes to their two most preferred alternatives and one vote to their least preferred alternative

BOUNDED RATIONALITY an assumption of satisficing behavior instead of strict rationality

BPC (BELIEFS, PREFERENCES, AND CONSTRAINTS) MODEL a model with only three assumptions: completeness, transitivity, and independence

BRINKMANSHIP a game between two players in which one player creates a threatening situation that forces the other player into a decision that is not rational

CARDINAL UTILITY preferences based on the numerical ordering of alternatives

CAUSALITY the statistical impact of one variable on another variable

CHAIN-STORE PARADOX a monopoly's decision concerning whether to fight a potential entrant

CHANCE MOVE a random move in a game that is determined by a probability distribution

CHEAP TALK in a signaling game, a message that does not impact the payoff of the sender but can impact the payoff of the receiver

CHICKEN GAME a game that illustrates brinkmanship

COMMITMENT DEVICE a device that commits a player to a certain course of action

COMMON KNOWLEDGE ASSUMPTION the assumption that players in a game know everything about the structure of the game

COMMON RATIO PROBLEM a situation in which one set of lotteries is reduced to the same probability to produce a second lottery

COMMON-POOL RESOURCE PROBLEM (CPR) a problem of public goods in terms of sharing and exploiting natural resources

COMPETITIVE GAMES games of conflict

COMPLETE (FULL) INFORMATION a state in which players know all strategies and payoffs of other players

COMPOUND LOTTERY a gamble in which a simple lottery is played first in order to win the chance to play in another simple lottery for which prizes are awarded

CONCAVE UTILITY FUNCTION an upward-sloping utility function that represents risk-averse or risk-hating behavior

CONDITIONAL PROBABILITY the probability of one event (or action) given the probability of another event or action

CONDORCET PARADOX a voting paradox in which three voters have preferences over three alternatives but the social ordering is intransitive

CONDORCET WINNER an alternative that beats all others in a pairwise comparison

CONFOUNDING FACTORS factors that interfere with the interpretation of measurements

CONNECTIVITY an assumption that specifies that two alternatives can be put in order such that if $A > B$, then $B > A$

CONSTANT-SUM GAMES games in which the payoffs to players do not sum to zero

CONTINUITY to use numbers to represent preferences we must also assume that outcomes behave like numbers. Numbers can be arranged from 1, 2, 3 so outcomes must also behave like numbers and be arranged from best outcome (1), the middle outcome (2) and the worse outcome (3). Continuity ensures that an identifiable middle exists that can align outcomes with the numbers

CONVEX UTILITY FUNCTION a downward-sloping utility function that represents risk-acceptance or risk-loving behavior

COOPERATIVE GAMES games in which players form coalitions and transfer utility among members

COURNOT BEST-RESPONSE DYNAMICS a learning model in which learning is derived from the last action of an opponent

CREDIBLE THREAT a threat that can be carried out

CROWDING OUT replacing some emotion or motivation with another emotion or motivation

CULTURAL LEARNING a learning model in which people learn based on the social norms that are prevalent within their environment

CUMULATIVE VOTING a voting system in which voters can cast two votes for one alternative

DATA-GENERATING PROCESS (DGP) the process of collecting data

DECEPTION the method of giving subjects misleading information during an experiment

DEGENERATE STRATEGY a strategy produced by the conversion of a mixed strategy to a pure strategy by assigning one pure strategy to probability 1 and every other strategy to probability 0

DEPENDENT VARIABLE the variable for which a researcher is trying to explain variations

DICTATOR GAME a game in which one player offers an amount of money to another player and if the second player rejects the offer, then the first player keeps the total amount of money

DOMINANT SOLVABLE EQUILIBRIUM an equilibrium that is found by eliminating strategies that are dominated by other strategies

DOMINANT STRATEGY a strategy for which all elements or payoffs are higher than the elements of another strategy

DOMINATE STRATEGY a strategy for which all elements or payoffs are higher than all of the elements of all other strategies

DOUBLE-BLIND EXPERIMENT an experiment in which subjects' identities are hidden from both other subjects and the researcher

ECOLOGICAL VALIDITY the extent to which an experiment matches the real-world environment it is attempting to study

EMPATHY caring for others

EMPIRICAL RESEARCH research that seeks to attain knowledge by direct observation and quantitative or qualitative analysis

ENDOGENOUS originating or deriving from within a model or system

ENDOWMENT an initial sum of cash given to subjects to spend during an experiment

ENTRY DETERRENCE a strategy of fighting a player's potential entrance to a particular market

ENVIRONMENT the experimental world that is created by the assumptions of an experimental design

EQUILIBRIUM a solution to a game in which all players' beliefs, strategies, actions, and payoffs are fixed or unchanging as a result of the assumptions of the game

EQUITY-AVERSE (OR INEQUITY-AVERSE) SOCIAL PREFERENCES a player's preferences that are concerned with the equality of his or her own payoff in relation to another person's payoff

EVOLUTIONARY GAME THEORY the biological study of game theory

EVOLUTIONARY LEARNING a learning model in which players are not searching for equilibrium strategies but are rather "born" in equilibrium and innately know the equilibrium strategies

EXOGENOUS originating or deriving from outside of a model or system

EXPECTED UTILITY what a player expects to win given the probabilities associated with each cardinal utility value

EXPERIENCE-WEIGHTED ATTRACTION a learning model that integrates rule learning and belief learning

EXPERIMENT a researcher's intentional manipulation of the data-generating process

EXPERIMENTAL DESIGN the procedures and instructions that are used to create an experiment

EXPLICIT RANDOMIZATION a device that allows for randomization, like a roulette wheel or shuffling cards

EXTENSIVE FORM GAME a game displayed in a tree diagram

EXTERNAL VALIDITY whether a result is valid across different datasets

FAIRNESS issues related to equal distribution of players' payoffs

FIELD DATA (OBSERVATIONAL DATA) data that occur naturally outside of the laboratory

FIELD EXPERIMENT an experiment in which a researcher attempts to control and measure some aspect of a naturally occurring event or situation

FINANCIAL INCENTIVE the money paid to a subject to participate in an experiment

FINITE REPEATED GAME a stage game played a fixed number of times

FIRST MOVER'S ADVANTAGE the payoff advantage that belongs to the first player in an extensive form game

FLAT FEE the money paid to a subject to participate in an experiment that is not impacted by the player's performance during the experiment

FOCAL POINT (SCHELLING POINT) a psychologically prominent outcome that is used in the absence of communication and seems to be a natural outcome

FRAMING the contextual way in which a state of nature or an environment is presented to players or subjects

FREE-RIDER a person who does not contribute to a public good but receives the benefits of the good

FULL (COMPLETE) INFORMATION a situation in which players know the strategies and payoffs of other players

GAME THEORY a theory that assumes rationality and examines the strategic interaction of people's decisions and the resulting outcomes

GENERIC ENVIRONMENT an environment that lacks any contextual elements or situations

HANDICAP COSTS the difference in signaling costs that a low type has to pay to be considered a high type

HAWK-DOVE GAME a variant of the chicken game

IMITATION LEARNING a learning model in which a player observes what successful players do and imitates them

IMPERFECT INFORMATION a situation in which a player does not know all the moves (or actions) in a game

INCOMPLETE INFORMATION a situation in which a player lacks knowledge about the strategies and payoffs of another player

INCOMPLETE INFORMATION EXTENSIVE FORM GAMES dynamic games in which a player has private information not observed by other players that dictates the play of the game

INDEPENDENCE the idea that if any part of two lotteries is the same, then what determines a preference between those two lotteries is the part of them that is different

INDEPENDENT VARIABLE the variable that is used to explain variation in the dependent variable

INDIVIDUAL DECISION THEORY a theory constructed on the observation of one player choosing among alternatives

INDUCED VALUE THEORY the idea that payments in an experiment should be salient and based on players' performance during the experiment

INFINITE REPEATED GAME (INFINITE HORIZON) a repeated game in which the ending constraint is not known with certainty

INFORMATION CASCADE a sequential move game in which some players free-ride off the information of other players and cause them to disregard their private information in favor of public information

INFORMATION SET a dashed line in an extensive form game that connects two or more nodes and indicates that a player does not know the node at which he or she is located

INSTITUTION the rules and procedures that govern decision-making behavior

INTERPERSONAL COMPARISONS OF UTILITY comparisons of utility among two or more individuals that cannot be made using ordinal utility, but can be made using cardinal utility

INTRANSITIVITY three alternatives that are not in a transitive ordering such that A > B > C and C > A

JOINTNESS OF SUPPLY shared ownership of a public good in which the supply of the good is inexhaustible

LAB-IN-FIELD EXPERIMENTS field experiments that use portable technology to implement computer laboratory controls

LABORATORY EXPERIMENT an experiment that is conducted in a single location to implement controls

LEARNING the way in which players incorporate new data (or information) into their belief systems

LOTTERY a gamble among alternatives or the probability of two or more events occurring (see simple lottery)

LYING misrepresentation of the true state of nature

MEDIAN VOTER THEOREM the idea that in a two-candidate election that takes place over a single dimension in which voters have single-peaked preferences that are known, candidates' strategies will converge to the preference of the median voter

MESSAGE in a signaling game, payoff-specific information concerning the state of nature

MINIMAX STRATEGY a strategy in which a player selects the worst available alternatives for a given strategy and then selects the best of the worst as a security level

MIXED STRATEGY a strategy in which player preferences are represented by a probability distribution over a strategy set

MIXED STRATEGY EQUILIBRIUM an equilibrium for which players' optimal mixed strategies are a best response to another player's mixed strategy

MODEL an abstract representation of some real-world situation or event

MULTIPLE EQUILIBRIA a game theory model that has more than one unique solution

NASH EQUILIBRIUM an outcome in which no player has an incentive to deviate from a selected strategy

NEUROECONOMICS the joining of neuroscience (the study of brain processes) with economics

NODE a decision point in an extensive form game

NONCREDIBLE THREAT a threat that cannot be carried out according to the rules of rationality

NONCOMPETITIVE GAME a game without conflict

NONCOOPERATIVE GAME a game in which players cannot form binding coalitions

NONEXCLUDABLE GOOD a good with benefits that everyone can enjoy for free

NONPERVERSE SELECTION RULE a rule that states that a player will always select the more preferred alternative

NORMAL FORM GAME a game displayed in tabular form

OBSERVATIONAL (FIELD) DATA data that occur naturally outside the laboratory

ONE-SHOT GAME (EXPERIMENT) a stage game that is only played once.

ORDINAL UTILITY a utility function for which alternatives are ranked on an ordinal scale

OTHER-REGARDING PREFERENCES a state in which a person is concerned with his or her own payoff, but is also concerned with another person's payoff

PARETO OPTIMALITY given a worse outcome for two players, an outcome that is better for both players

PAYOFF the rewards a player receives, which can be money or anything that increases or decreases a person's welfare

PAYOFF-INTERDEPENDENT PREFERENCES a state in which subjects have preferences not only for money, but also for other matters not controlled for in the experiment

PERFECT INFORMATION a situation in which a player knows the moves (actions) of other players, including a complete history of moves

PERFECT RECALL the inability of a player to forget how he or she has moved

PERFORMANCE-BASED FINANCIAL INCENTIVES money paid to a subject during an experiment in which the subject's actions during the experiment determine how much money he or she receives

PERSUASION the ability of one player to influence and alter the beliefs of another player

PLAYER a person in a game theory model, usually represented by a mathematical function

PLURALITY VOTING RULE a voting rule that states that the alternative with the most votes wins the election

POLITICAL ECONOMY EXPERIMENTS experiments conducted by biologists, economists, neuro-economists, political scientists, psychologists, and sociologists to test rational choice theories

POOLING EQUILIBRIUM in a signaling game, an equilibrium in which senders of both types choose the same strategy and prevent the receiver from knowing the sender's type

POSITRON EMISSION TOMOGRAPHY (PET) imaging of a subject's brain

POSTERIOR PROBABILITIES beliefs that have been updated with new information

PREDATORY PRICING the lowering of prices to drive out a competitor

PRIORS players' established beliefs, expressed as a probability number like 50%

PRISONER'S DILEMMA a game in which players have a dominant strategy to not cooperate and the selection of that strategy leads to a worse outcome than if they cooperated

PRIVATE INFORMATION information to which one player is privy and another player is not

PRO-SOCIAL BEHAVIOR voluntary actions on the part of an individual that are intended to help or benefit one or more individuals

PROSPECT THEORY the idea that people gain utility with a concave function but lose utility with a convex function

PUBLIC (COLLECTIVE) GOOD a good that is produced by public money that has the properties of nonexcludability and jointness of supply

PUBLIC INFORMATION information that is announced to all players in a game theory model or experiment

PURE STRATEGY the selection of only one strategy

PURE THEORY a theory that is valued for its results and for which there is no concern about testing those results

RABIN'S FAIRNESS UTILITY FUNCTION a function that allows players to derive utility from being nice to players who are perceived to be nice and being mean to players who are perceived to be mean

QUANTAL RESPONSE EQUILIBRIUM (QRE) a statistical version of a Nash equilibrium that allows for a stochastic element to choices when analyzing experimental data

QUORUM-BUSTING an attempt by a minority party to defeat legislation favored by a majority party by not showing up to a legislature, so that a quorum is not present and an official vote cannot be taken

RANDOM GENERATING DEVICE a device that selects random strategies

RATIONAL CHOICE THEORY a theory that uses expected utility to mathematically model preference choice

RATIONALITY the idea that people can order choices and will always select the best possible choice

RECEIVER in a signaling game, the person who receives a message from the sender

RECIPROCITY NORM the idea that a person's behavior is a function of another person's behavior in two or more periods

REDUCTION OF COMPOUND LOTTERIES the idea that complex lotteries can be reduced to simple lotteries

REGRET THEORY a theory that states that a person experiences an added loss in utility when he or she regrets a decision

REPEATED TRIAL[1] a single trial that is repeated more than one time

REPEATED GAME a stage game played more than one time

REPLICATION OF DATA the ability to reproduce a dataset

RISK ACCEPTANCE a person's tendency to embrace risky situations

RISK AVERSION a person's tendency to shy away from risky situations

RISK DOMINANT EQUILIBRIUM an equilibrium with the smallest risk factor

ROUTE CHOICE GAME a repeated prisoner's dilemma game that examines traffic congestion

RULE LEARNING a learning model in which players learn a set of plausible limiting decision rules that allow them to choose strategies

SCRIPT a contextual description of a situation that is used as a stimulus to motivate subjects in an experiment

SECOND MOVER'S ADVANTAGE the payoff advantage that a second player in an extensive form game enjoys in game play

SELF-REGARDING PREFERENCES a state in which a person is only concerned with his or her payoff; also, the standard assumption of rationality

SEMI-POOLING (OR SEPARATING) EQUILIBRIUM in a signaling game, an equilibrium that includes a mixed strategy

SENDER in a signaling game, the player who sends a message

SEPARATING EQUILIBRIUM in a signaling game, an equilibrium in which each sender type chooses a different strategy, given his or her type, such that the receiver knows the type of the sender

SEQUENTIAL EQUILIBRIUM an equilibrium for extensive form games that assumes simple probability beliefs and sequential rationality

SEQUENTIAL RATIONALITY a condition that ensures that a player will select the alternative with the highest outcome at decision nodes

SEQUENTIAL VOTING (ELECTIONS) elections that occur in stages, with the final vote tally being the accumulated vote total from all stages

SHOW-UP FEE money paid to a subject for simply showing up on time for an experiment

SIGNAL a message sent to a receiver

SIGNALING COST the cost to a player to send a message

SIGNALING GAME a game in which a sender has private information about the state of the world and sends a message to a receiver, who is not privy to this private information but must take some action

SIMPLE LOTTERY a gamble between two alternatives such that the probabilities sum to one

SINCERE VOTING a player's tendency to always vote for the alternative that is ranked first in his or her list of alternatives

SINGLE-BLIND EXPERIMENT an experiment in which subjects' identities are hidden from each other but not from the researcher

SINGLE-PEAKED UTILITY FUNCTION a graphical utility function that has a single peak at a player's ideal point

SINGLETON a single node in an extensive form game

SOCIAL CHOICE THEORY a theory that examines how aggregation of preferences results in social outcomes

SOCIAL PREFERENCES a player's concern for another player's payoff

SPATIAL THEORY OF ELECTIONS AND COMMITTEES a theory that examines majority-rule outcomes for situations in which players have spatial preferences

STAG HUNT a game that illustrates assurance problems or trust games

STAGE GAME a normal form game or an initial game which may be played a repeated number of times

STATE OF NATURE two or more situations governed by a probability distribution for which payoffs differ

STRAIGHT VOTING a voting system in which voters can cast one vote for up to two alternatives

STRATEGIC IGNORANCE the theory that given a choice of information, people prefer to avoid information that could impact their beliefs and, in turn, their behavior

STRATEGIC VOTING a method of voting for a least preferred alternative in order to achieve a more preferred outcome

STRATEGY a choice of an action given another player's choice of an action

STRATEGY MAPPING a player's map of all of his possible strategy contingencies based on all possible strategy contingencies of an opponent

SUBGAME part of an extensive form game that begins at a single node and, when detached from the original game, can also be treated as a game

SUBGAME PERFECT EQUILIBRIUM an equilibrium that eliminates Nash equilibria that is supported by a noncredible threat

SUBJECT a participant in an experiment

TERMINAL HISTORIES the ending nodes in an extensive form game for which payoffs are usually specified

TRANSITIVITY the ordering of three alternatives such that if A > B > C then A > C

TREATMENT EFFECT the mean behavior in one treatment minus the mean behavior in another treatment

TREATMENT (OR TREATMENT VARIABLE) a manipulation in an experiment that is usually varied

TRUSTWORTHINESS a person's trust that some payoff-relevant information is true or that another person will take a certain action

TWO-DIMENSIONAL SPATIAL UTILITY FUNCTION a graphical utility function in which a player has preferences on two dimensions and an ideal point is represented by a dot in the two dimensions

ULTIMATUM GAME a game in which Player 1 offers Player 2 an amount of money and if Player 2 rejects the offer, the game is over and both players receive nothing

UNCERTAINTY PRINCIPLE the idea that certain parameters in a model will never be known, so some randomness in the form of a probability distribution can predict the unknown parameter and the model more precisely

URN a device like a jar that a person can use to make random picks

UTILITY the satisfaction that a person gains from a particular alternative

UTILITY FUNCTION a mathematical function that maps numbers or ordered outcomes to utility for a player

VARIABLE a concept that can be measured

VON NEUMANN-MORGENSTERN (VNM) UTILITY FUNCTION a utility function that involves lotteries

WEAK CONSISTENCY OF BELIEFS the idea that for every information set on the equilibrium path that is reached with positive probability, beliefs are defined by Bayes' rule

WEAKLY DOMINANT STRATEGY a strategy for which all elements are at least as good as another strategy's elements

WEIGHTED FICTITIOUS PLAY a learning model based on the past history of another player in which weight is given to more current actions

WITHIN-SUBJECT DESIGN an experimental design with at least two treatments in which subjects are exposed to both treatments (usually randomly during successive trials)

ZERO-SUM GAME a game in which players' payoffs sum to zero

NOTES

1. I would like to note that this term means the same thing as repeated games. Although the terms *single trial* and *stage game* appear to refer to different things, they are the same. When the term *repeated* is used, it simply means that the game we are considering is played more than one time, regardless of how it is classified.

INDEX